Springer Texts in Statistics

Springer Texts in Statistics

Alfred	Elements of Statistics for the Life and Social Sciences
Berger	An Introduction to Probability and Stochastic Processes
Blom	Probability and Statistics: Theory and Applications
Chow and Teicher	Probability Theory: Independence, Interchangeability, Martingales Second Edition
Christensen	Plane Answers to Complex Questions: The Theory of Linear Models
Christensen	Linear Models for Multivariate, Time Series, and Spatial Data
Christensen	Log-Linear Models
du Toit, Steyn and Stumpf	Graphical Exploratory Data Analysis
Finkelstein and Levin	Statistics for Lawyers
Jobson	Applied Multivariate Data Analysis, Volume I: Regression and Experimental Design
Jobson	Applied Multivariate Data Analysis, Volume II: Categorical and Multivariate Methods
Kalbfleisch	Probability and Statistical Inference: Volume 1: Probability Second Edition
Kalbfleisch	Probability and Statistical Inference: Volume 2: Statistical Inference Second Edition

(continued after index)

Marc A. Berger

An Introduction to Probability and Stochastic Processes

With 19 Illustrations

Springer-Verlag
New York Berlin Heidelberg London Paris
Tokyo Hong Kong Barcelona Budapest

Marc A. Berger
School of Mathematics
Georgia Institute of Technology
Atlanta, GA 30332-0269
USA

Cover illustration: A zoom-in view of a fractal dragon set. This image is generated from a 2-map iterated function system. (See color plate 11.)

Mathematics Subject Classifications (1991): 60-01, 60E05, 60J10, 60J27, 60E10, 60F05, 60F15, 60G10, 47A35

Library of Congress Cataloging-in-Publication Data
Berger, Marc A., 1955–
An introduction to probability and stochastic processes / Marc A. Berger.
p. cm.
Includes bibliographical references and index.
ISBN 0-387-97784-8
1. Probabilities. 2. Stochastic processes. I. Title.
QA273.B48 1992
519.2—dc20 91-43019

Printed on acid-free paper.

Production managed by Henry Krell, manufacturing supervised by Vincent Scelta.
Typeset by Asco Trade Typesetting Ltd., Hong Kong.
Printed and bound by R.R. Donnelley & Sons, Harrisonburg, VA.
Printed in the United States of America.

9 8 7 6 5 4 3 2 1

ISBN 0-387-97784-8 Springer-Verlag New York Berlin Heidelberg
ISBN 3-540-97784-8 Springer-Verlag Berlin Heidelberg New York

In memory of Dr. Robert A. Dannels, a great man and a dear friend who always provided advice and encouragement at critical times—with his unique touch of humor.

Preface

These notes were written as a result of my having taught a "nonmeasure theoretic" course in probability and stochastic processes a few times at the Weizmann Institute in Israel. I have tried to follow two principles. The first is to prove things "probabilistically" whenever possible without recourse to other branches of mathematics and in a notation that is as "probabilistic" as possible. Thus, for example, the asymptotics of P^n for large n, where P is a stochastic matrix, is developed in Section V by using passage probabilities and hitting times rather than, say, pulling in Perron-Frobenius theory or spectral analysis. Similarly in Section II the joint normal distribution is studied through conditional expectation rather than quadratic forms.

The second principle I have tried to follow is to only prove results in their simple forms and to try to eliminate any minor technical computations from proofs, so as to expose the most important steps. Steps in proofs or derivations that involve algebra or basic calculus are not shown; only steps involving, say, the use of independence or a dominated convergence argument or an assumption in a theorem are displayed. For example, in proving inversion formulas for characteristic functions I omit steps involving evaluation of basic trigonometric integrals and display details only where use is made of Fubini's Theorem or the Dominated Convergence Theorem. Similarly in proving the Berry-Esseen bound I leave out steps where algebraic substitution and simplification are carried out. This was done to make the proofs more transparent and to not clutter up the arguments.

In many texts, generalized versions of the limit theorems are presented, and as a result the proofs get involved. My point in these notes is to show that the basic limit results can be proved concisely. For example, in proving the Strong LLN some authors will first show that a.s. convergence and convergence in probability are equivalent for sums of independent random variables. Confining myself to proving the basic Strong Law, I only use Kolmogorov's Maximal Inequality, like in Billingsley [4]; as a result the proof fits into two to three pages. Similarly many authors prove the Berry-Esseen bound for Lindeberg-Feller versions of the CLT. I confine myself to proving this bound for the basic CLT (in its original form) as a result of which the proof, taken from Feller [17], is rather quick.

The notes here are designed to fill a one-semester course. I have tried to make them complete and self-contained. With just a few exceptions every result needed is proved herein rather than quoted from outside. Among the exceptions are the Dominated Convergence Theorem, Fubini's Theorem, Abel's Lemma, and the fact that the Legendre-Fenchel transform of a strictly convex function is essentially differentiable. My feeling in these cases was that effort in these results would both sidetrack too much and push the notes out of the range of a single semester. I also did not prove the limit result for $P(t)$ as $t \to \infty$, for Markov pure jump processes. Because of the brevity of these notes it is important to supplement details about background and perspective. Thus, for example, before studying extremes and extremal distributions in Section III, where the Poisson distribution comes in, one should first discuss order statistics, where the binomial distribution comes in.

The material in these notes can be grouped into four categories. The first two sections deal with the theory of random variables and distributions. The third section covers the basic limit theorems. The fourth, fifth and sixth sections cover discrete and continuous time Markov processes. Finally, the last section covers products of random matrices and their application to generation of fractals. As mentioned above, these notes are intended to be an instructional guide, and as such the material contained herein was merged together from many different sources. Some of the presentation is my own, but most of it comes from the references listed. I drew most heavily from Billingsley [4], Breiman [5] and Feller [17] for the first three sections and from Hoel, Port and Stone [28], Karlin and Taylor [31] and Parzen [45] for the next three sections. Most of the collected exercises also come from these authors. (Incidentally they vary greatly in levels of difficulty.)

In the most concise way, my advice to anyone teaching "nonmeasure theoretic" probability at a serious level is to emulate Feller's marvelous volumes as much as possible. Much of my excitement about the field of probability theory stems from his writings.

Acknowledgments for Permissions

Springer-Verlag wishes to thank the publishers listed below for their copyright permission and endorsement to use their previously published material in this book. Their invaluable help in this matter has made the publication of this volume possible.

Exercises 10 in Chapter 2; 19, 25, 27, and 28 in Chapter 4; 8, 9, 10, 19, and 20 in Chapter 5; and 4, 7, 8, 9, 18, 19, and 20 in Chapter 6 have been reproduced with the kind permission of Academic Press Inc. from *A First Course in Stochastic Processes* (*Second Ed.*) by S. Karlin and H.M. Taylor.

Material in this book appeared in Sections 1.2 and 1.3 (pages 7–12) and Exercise 13 in Section 9.1 (page 306) in *A Course in Probability Theory* by K.L. Chung, and has been reproduced with the kind permission of Academic Press Inc. and K.L. Chung.

Exercises 2 and 3 in Chapter 2 has been reprinted with permission from *Probability, Classics in Applied Probability, Number* 7, by Leo Breiman. Copyright 1992 by the Society for Industrial and Applied Mathematics, Philadelphia, Pennsylvania. All rights reserved.

Exercises 5, 6, 7, and 8 in Chapter 1; 11 in Chapter 2; and 2, 3 and 4 in Chapter 3 have been reproduced with the kind permission of Chelsea Publishing Company from *The Theory of Probability* by B.V. Gnedenko.

Exercises 1, 2, 3, and 4 in Chapter 1; 1, 4, 5, 6, 7, 8, 12, 13, 14, 15, 16, 17, and 18 in Chapter 2; 1(a–i), 1(k), 7, 18 and 26 in Chapter 4; 1(a–i), 1(k), 3, 6, and 7 in Chapter 5; and 11, 12, 13, and 14 in Chapter 6 have been reproduced with the kind permission of E. Parzen from *Stochastic Processes* by E. Parzen. Also, material in this book appeared in Chapters 6 and 7 of *Stochastic Processes*.

Exercises 1(j), 1(l), 2, 3, 4, 5, 6, 11, 12, 13, 14, 15, 16, 17, 21, and 23 in Chapter 4; 1(j), 1(l), 2, 4, 5, 11, 13, 14, 17, and 18 in Chapter 5; and 1, 2, 5, 6, 15, and 16 in Chapter 6 have been reproduced with the kind permission of Houghton Mifflin Publishing Company from *Introduction to Stochastic Processes* by P.G. Hoel, S.C. Port and C.J. Stone. Also, material in Chapters 4–6 of this book appeared in Chapters 1–3 of Hoel, P.G., S.C. Port, and C.J. Stone, *Introduction to Stochastic Processes.* Copyright © 1972 by Houghton Mifflin Company. Used with permission.

Material in this book appeared on pages 219–222 of *Probability Theory I (4th Edition)* by M. Loeve, and has been reproduced with the kind permission of Springer-Verlag.

Material in this book appeared in Sections 3 and 4 of *Large Deviations and Applications* by S.R.S. Varadhan. Reprinted with permission from the CBMS-NSF Regional Conference Series in Applied Mathematics, Number 46. Copyright 1984 by the Society for Industrial and Applied Mathematics, Philadelphia, Pennsylvania. All rights reserved.

Material in this book appeared in Example XV.2(j) on page 380, and on pages 452–453 in *An Introduction to Probability Theory and Its Applications, Volume I (3rd ed.)* by W. Feller, © 1968. Reprinted by permission of John Wiley & Sons, Inc.

Material in this book appeared in Section 22 (pages 248–251) of *Probability and Measure* by P. Billingsley, © 1979. Reprinted by permission of John Wiley & Sons, Inc.

Contents

Preface vii

Acknowledgments for Permissions ix

I. Univariate Random Variables 1
- Discrete Random Variables 1
- Properties of Expectation 2
- Properties of Characteristic Functions 6
- Basic Distributions 8
- Absolutely Continuous Random Variables 11
- Basic Distributions 16
- Distribution Functions 20
- Computer Generation of Random Variables 23
- Exercises 24

II. Multivariate Random Variables 27
- Joint Random Variables 27
- Conditional Expectation 33
- Orthogonal Projections 36
- Joint Normal Distribution 38
- Multi-Dimensional Distribution Functions 39
- Exercises 42

III. Limit Laws 45
- Law of Large Numbers 45
- Weak Convergence 48
- Bochner's Theorem 58

Extremes 60
Extremal Distributions 65
Large Deviations 69
Exercises 76

IV. Markov Chains—Passage Phenomena 78
First Notions and Results 78
Limiting Diffusions 89
Branching Chains 91
Queueing Chains 93
Exercises 96

V. Markov Chains—Stationary Distributions and Steady State 101
Stationary Distributions 101
Geometric Ergodicity 110
Examples 111
Exercises 117

VI. Markov Jump Processes 121
Pure Jump Processes 121
Poisson Process 125
Birth and Death Process 129
Exercises 135

VII. Ergodic Theory with an Application to Fractals 139
Ergodic Theorems 139
Subadditive Ergodic Theorem 143
Products of Random Matrices 146
Oseledec's Theorem 149
Fractals 156
Bibliographical Comments 166
Exercises 168

References 173

Solutions (Sections I–V) 177

Index 201

Section I
Univariate Random Variables

Discrete Random Variables

These are real-valued functions X defined on a probability space, taking on a finite or countably infinite number of values $\{x_1, x_2, \ldots\}$. They can be described by a *discrete density function*

$$p_X(x) = \mathbb{P}(X = x).$$

Such a density function has the following properties:

$$\text{(i)}\ p_X(x) \geq 0, \qquad x \in \mathbb{R}$$

$$\text{(ii)}\ \{x : p_X(x) \neq 0\} \text{ is a finite or countably infinite set}$$

$$\text{(iii)}\ \sum_x p_X(x) = 1.$$

Typically discrete random variables are integer-valued. Random variables describe measured outcomes from experiments in which randomness (or nondeterminism) contributes.

We say that X has *finite expectation* if

$$\sum_x |x| p_X(x) < \infty.$$

In this case we define its *expectation* EX to be

$$EX = \sum_x x p_X(x). \tag{1}$$

Suppose f is a real-valued function defined on $\mathbb{R}$. We would like to consider the random variable $Y = f(X)$. The possible values for Y are $y_i = f(x_i)$ and

$$p_Y(y) = \mathbb{P}(X \in F^{-1}(y)) = \sum_{x \in f^{-1}(y)} p_X(x). \tag{2}$$

Here we allow for the possibility that f may not be one to one. From this it follows that

$$\sum_y |y| p_Y(y) = \sum_y |y| \sum_{x \in f^{-1}(y)} p_X(x) = \sum_x |f(x)| p_X(x).$$

We conclude from this that Y has finite expectation if and only if $\sum_x |f(x)| p_X(x) < \infty$; if this holds, then by a similar calculation

$$\mathbf{E}f(X) = \sum_x f(x) p_X(x). \tag{3}$$

Observe that (3) is consistent in the following sense. Suppose f and g are two functions for which $f(X) = g(X)$. This happens when $f(x) = g(x)$ for all x at which $p_X(x) > 0$. Then it follows from (3) that $\mathbf{E}f(X) = \mathbf{E}g(X)$. We shall have occasion to deal with *complex*-valued functions f defined on $\mathbb{R}$. In this case we say that $f(X)$ has finite expectation if each of $f_1(X)$ and $f_2(X)$ have finite expectation, $f_1 = \mathrm{Re}(f)$ being the real part of f and $f_2 = \mathrm{Im}(f)$ being the imaginary part of f. Since

$$|f_1|, |f_2| \leq |f| \leq |f_1| + |f_2|$$

this is equivalent to the condition $\sum_x |f(x)| p_X(x) < \infty$ ($|f|$ and $|f(x)|$ here denote the modulus of the complex entities). If this condition holds, then we define

$$\mathbf{E}f(X) = \mathbf{E}f_1(X) + i\mathbf{E}f_2(X) = \sum_x f(x) p_X(x). \tag{4}$$

Properties of Expectation

(P_1) *If* $\mathrm{f_i}$ *are complex-valued functions defined on* $\mathbb{R}$ *for which* $\mathrm{f_i(X)}$ *have finite expectation, and if* $\mathrm{a_i}$ *are (complex) constants,* $1 \leq \mathrm{i} \leq \mathrm{n}$, *then* $\sum_{\mathrm{i}=1}^{\mathrm{n}} \mathrm{a_i f_i(X)}$ *has finite expectation and*

$$\mathbf{E}\left[\sum_{i=1}^{n} a_i f_i(X)\right] = \sum_{i=1}^{n} a_i \mathbf{E} f_i(X). \tag{5}$$

In particular if X *has finite expectation and if* a *and* b *are constants then* $\mathrm{aX + b}$ *has finite expectation and* $\mathbf{E}(\mathrm{aX + b}) = \mathrm{a}\mathbf{E}\mathrm{X} + \mathrm{b}$. *Also,* $\mathbf{E}\mathrm{b} = \mathrm{b}$ (*thinking of* b *on the left as a constant random variable*).

(P_2) *If* X *has finite expectation and* $\mathrm{X} \geq 0$ *then* $\mathbf{E}\mathrm{X} \geq 0$. *Moreover* $\mathbf{E}\mathrm{X} = 0$ *in this case if and only if* $\mathrm{X} = 0$. *In particular if* f *and* g *are real-valued functions defined on* $\mathbb{R}$ *for which* f(X) *and* g(X) *have finite expectation, and if* $\mathrm{f(X)} \leq \mathrm{g(X)}$[1], *then* $\mathbf{E}\mathrm{f(X)} \leq \mathbf{E}\mathrm{g(X)}$, *with equality if and only if* $\mathrm{f(X) = g(X)}$. *Also, if* X *has finite expectation then* $|\mathbf{E}\mathrm{X}| \leq \mathbf{E}|\mathrm{X}|$.

[1] This means that $f(x) \leq g(x)$ for all x at which $p_X(x) > 0$.

(P_3) *If* $m_1 \le X \le m_2$ *for some constants* m_1 *and* m_2, *then* X *has finite expectation and* $m_1 \le \mathbf{E}X \le m_2$. *In particular if* $|X| \le \varepsilon$ *for some constant* $\varepsilon > 0$ *then* $|\mathbf{E}X| \le \varepsilon$.

(P_4) *If* X *is nonnegative integer-valued, then* X *has finite expectation if and only if the series* $\sum_{x=1}^{\infty} \mathbb{P}(X \ge x)$ *converges. If this series does converge then its sum is* **EX**.

(P_5) **Jensen's Inequality.** *If* X *has finite expectation and if* f *is a convex real-valued function defined on* $\mathbb{R}$ *for which* f(X) *has finite expectation then*

$$f(\mathbf{E}X) \le \mathbf{E}f(X). \tag{6}$$

PROOFS.
(P_1) First observe that

$$\sum_x \left| \sum_{i=1}^n a_i f_i(x) \right| p_X(x) \le \sum_x \sum_{i=1}^n |a_i f_i(x)| p_X(x)$$
$$= \sum_{i=1}^n |a_i| \sum_x |f_i(x)| p_X(x) < \infty.$$

so that $\sum_{i=1}^n a_i f_i(X)$ has finite expectation. Thus by (3)

$$\mathbf{E}\left[\sum_{i=1}^n a_i f_i(X)\right] = \sum_x \left[\sum_{i=1}^n a_i f_i(x)\right] p_X(x)$$
$$= \sum_{i=1}^n a_i \sum_x f_i(x) p_X(x) = \sum_{i=1}^n a_i \mathbf{E} f_i(X). \qquad \square$$

(P_2) If $X \ge 0$ then $p_X(x) = 0$ for $x < 0$. Thus

$$\mathbf{E}X = \sum_x x p_X(x) = \sum_{x \ge 0} x p_X(x) \ge 0.$$

Furthermore, equality holds here if and only if $p_X(x) = 0$ for $x > 0$, in which case $p_X(0) = 1$. Using (P_1) and applying this result here to the random variable $g(X) - f(X)$, and then specializing to the choices $f(x) = \pm x$, $g(x) = |x|$ leads to the other conclusions in (P_2). $\square$

(P_3) Let $m = \max(|m_1|, |m_2|)$. Since $p_X(x) = 0$ if $x \notin [m_1, m_2]$ it follows in particular that $p_X(x) = 0$ if $|x| > m$. Thus

$$\sum_x |x| p_X(x) = \sum_{|x| \le m} |x| p_X(x) \le \sum_{|x| \le m} m p_X(x) = m.$$

From this we see that X has finite expectation. The fact that $m_1 \le \mathbf{E}X \le m_2$ follows from (P_2) and (the very last part of) (P_1). $\square$

(P_4)
$\sum_{x=1}^{\infty} x p_X(x) = \sum_{x=1}^{\infty} p_X(x) \sum_{y=1}^{x} 1 = \sum_{y=1}^{\infty} \sum_{x=y}^{\infty} p_X(x) = \sum_{y=1}^{\infty} \mathbb{P}(X \ge y)$. $\square$

(P_5) We need to show that if $p_i > 0$, $\sum p_i = 1$ then

$$f\left(\sum_i p_i x_i\right) \leq \sum_i p_i f(x_i).$$

If there are only a finite number of p_is, then this inequality follows directly from the convexity of f. Otherwise, observe that for any n

$$f\left(\frac{\sum_{i=1}^n p_i x_i}{\sum_{i=1}^n p_i}\right) \leq \frac{\sum_{i=1}^n p_i f(x_i)}{\sum_{i=1}^n p_i}.$$

Since convex functions are necessarily continuous, we can take limits as $n \to \infty$ and arrive at the desired result. □

Let X be a random variable such that X^2 has a finite expectation. Since $|x| \leq x^2 + 1$ it is easily seen that X itself also has a finite expectation. We define the *variance* of X to be

$$\operatorname{Var} X = \mathbf{E}(X - \mathbf{E}X)^2. \tag{7}$$

This is always a nonnegative number, and we write $\operatorname{Var} X = \sigma_X^2$ and refer to σ_X as the *standard deviation* of X. Since $\mathbf{E}X = \mu$ is a constant it follows from (P_1) that the variance is also given by the expression

$$\operatorname{Var} X = \mathbf{E}X^2 - \mu^2. \tag{8}$$

One consequence of this is that $\mathbf{E}X^2 \geq (\mathbf{E}X)^2$, for any random variable X such that X^2 has finite expectation. For any constant a

$$\mathbf{E}(X - a)^2 = \mathbf{E}(X - \mu)^2 + (\mu - a)^2, \tag{9}$$

so that $\operatorname{Var} X$ is the minimum value of $\mathbf{E}(X - a)^2$, that minimum being realized at $a = \mu$. This allows us to interpret $\mathbf{E}X$ as the best constant approximation to X, in the least squares sense.

Let $r \geq 0$ be an integer. We say that X has a *moment of order* r if X^r has finite expectation, and in that case we define the rth *moment* of X as $\mathbf{E}X^r$. If X has a moment of order r, then it has a moment of order k for all $k \leq r$, since $|x|^k \leq |x|^r + 1$. If X has a moment of order r then $X - \mu$ has a moment of order r, by (P_1), where $\mu = \mathbf{E}X$; it is referred to as the rth *central moment* of X. Thus the first centered moment of X is always zero (whenever X has a finite expectation), and the second centered moment of X is its variance (whenever X^2 has a finite expectation).

The *characteristic function* of a random variable X is defined as

$$\varphi_X(u) = \mathbf{E}e^{iuX}, \qquad -\infty < u < \infty. \tag{10}$$

Since $|e^{iuX}| = 1$ it follows that e^{iuX} always has finite expectation, for any value of u, and thus φ_X is well-defined. Observe that

$$\varphi_X(u) = \sum_x e^{iux} p_X(x). \tag{11}$$

When X is integer-valued this is simply the Fourier series for φ_X. It is special in that its coefficients are nonnegative numbers summing to one. In particular since $\sum_x p_X^2(x) < \infty$, $\varphi \in L^2(-\pi, \pi)$ and we can recover the distribution of X from φ via its Fourier coefficients

$$p_X(x) = \frac{1}{2\pi}\int_{-\pi}^{\pi} e^{-iux}\varphi_X(u)\,du; \qquad x = 0, \pm 1, \pm 2, \ldots. \tag{12}$$

Slightly more generally, if X takes on the values $\{kd : k = 0, \pm 1, \pm 2, \ldots\}$ then the same argument shows that

$$p_X(x) = \frac{d}{2\pi}\int_{-\pi/d}^{\pi/d} e^{-iux}\varphi_X(u)\,du; \qquad x = 0, \pm d, \pm 2d, \ldots. \tag{13}$$

In general, if the range of X is not assumed to lie inside some lattice, then

$$p_X(x) = \lim_{N\to\infty}\frac{1}{2N}\int_{-N}^{N} e^{-iux}\varphi_X(u)\,du. \tag{14}$$

The proof of (14) depends on the following results from real analysis.

I. **(Dominated Convergence Theorem)** Let f_n and f be real-valued functions defined on $\mathbb{R}$ $(n = 1, 2, \ldots)$, and suppose that for each x

$$\lim_{n\to\infty} f_n(x) = f(x).$$

If there exists a real-valued function g for which $g(X)$ has finite expectation, satisfying $|f_n| \le g$ $(n = 1, 2, \ldots)$, then

$$\lim_{n\to\infty} \mathbf{E}f_n(X) = \mathbf{E}f(X).$$

II. **(Fubini's Theorem)** Let f be a complex-valued function of two variables, continuous in the second (everywhere). If there exists a real-valued function g for which $g(X)$ has finite expectation, satisfying

$$\int_{-\infty}^{\infty} |f(x, u)|\,du \le g(x)$$

for each x, then

$$\int_{-\infty}^{\infty} \mathbf{E}f(X, u)\,du = \mathbf{E}\int_{-\infty}^{\infty} f(X, u)\,du.$$

PROOF OF (14)

$$\begin{aligned}\frac{1}{2N}\int_{-N}^{N} e^{-iux}\varphi_X(u)\,du &= \frac{1}{2N}\int_{-N}^{N} \mathbf{E}e^{iu(X-x)}\,du\\ &= \frac{1}{2N}\mathbf{E}\int_{-N}^{N} e^{iu(X-x)}\,du \qquad \text{(by (II) with } g \equiv 1)\\ &= \mathbf{E}\frac{\sin N(X-x)}{N(X-x)}.\end{aligned}$$

Since $\lim_{N\to\infty} \frac{\sin N(y-x)}{N(y-x)} = I_{\{x\}}(y)$ and since $\left|\frac{\sin t}{t}\right| \le 1$ for all t, it follows from Result I that

$$\lim_{N\to\infty} \mathbf{E}\frac{\sin N(X-x)}{N(X-x)} = \mathbf{E}I_{\{x\}}(X) = p_X(x). \qquad \square$$

Properties of Characteristic Functions

(P_1) $\varphi_X(0) = 1$.

(P_2) $|\varphi_X| \le 1$.

(P_3) $\varphi_X(-u) = \overline{\varphi_X(u)}$.

(P_4) *If* X *has a moment of order* k, *then* φ_X *is* k *times differentiable at* $u = 0$, *and*

$$\mathbf{E}X^k = \frac{1}{i^k}\frac{d^k}{du^k}\varphi_X(0). \tag{15}$$

(P_5) φ_X *determines the distribution of* X *uniquely.*

(P_6) φ_X *is uniformly continuous on* $\mathbb{R}$.

(P_7) φ_X *is positive-semidefinite in the sense that*

$$\sum_{j=1}^{n}\sum_{k=1}^{n} \varphi_X(u_j - u_k)\xi_j\bar{\xi}_k \ge 0 \tag{16}$$

for any real numbers $u_1, u_2, \ldots, u_n$ *and any complex numbers* $\xi_1, \xi_2, \ldots, \xi_n$.

Proofs. Properties (P_1)–(P_3) are immediate and property (P_5) follows from (14). $\square$

(P_4) Use Taylor's Theorem with remainder on $\sin t$ and $\cos t$ to write

$$e^{iuX} = \sum_{j=0}^{k} \frac{(iuX)^j}{j!} + \frac{(iuX)^k}{k!}[\cos(\theta_1(X)uX) + i\sin(\theta_2(X)uX) - 1],$$

where $|\theta_1|, |\theta_2| \le 1$. For each x

$$\lim_{u\to 0} x^k[\cos(\theta_1(x)ux) + i\sin(\theta_2(x)ux) - 1] = 0.$$

Furthermore, for each u

$$|x^k[\cos(\theta_1(x)ux) + i\sin(\theta_2(x)ux) - 1| \leq 3|x|^k.$$

Result I from real analysis applies even if the index n is allowed to be real rather than integral. Using this version of it we conclude that

$$\varphi_X(u) = \sum_{j=0}^{k-1} \frac{(iu)^j}{j!}\mathbf{E}X^k + \frac{(iu)^k}{k!}[\mathbf{E}X^k + o(1)]. \qquad \square$$

$$(\mathrm{P}_6) \qquad |\varphi_X(u+h) - \varphi_X(u)| = |\mathbf{E}e^{iuX}(e^{ihX} - 1)| \leq \mathbf{E}|e^{ihX} - 1|.$$

Apply Result I as $h \to 0$ to conclude that

$$\lim_{h\to 0} \mathbf{E}|e^{ihX} - 1| = 0. \qquad \square$$

$$(\mathrm{P}_7) \qquad \sum_{j=1}^{n}\sum_{k=1}^{n} \varphi_X(u_j - u_k)\xi_j\bar{\xi}_k = \mathbf{E}\sum_{j=1}^{n}\sum_{k=1}^{n} e^{i(u_j-u_k)X}\xi_j\bar{\xi}_k = \mathbf{E}\left|\sum_{j=1}^{n} e^{iu_jX}\xi_j\right|^2 \geq 0. \qquad \square$$

If X has a moment of order k then we define its kth *cumulant* to be

$$\frac{1}{i^k}\frac{d^k}{du^k}\log \varphi_X(0).$$

The *moment generating function* $\psi_X(t)$ of a random variable X is defined by

$$\psi_X(t) = \mathbf{E}e^{tX}. \tag{17}$$

Its domain consists of all real numbers t such that e^{tX} has finite expectation, and it is easily seen that this turns out to be an interval containing $t = 0$. In order for ψ_X to exist for t in some neighborhood of zero, it is necessary that X have moments of all orders. These moments are then computable as

$$\mathbf{E}X^k = \frac{d^k}{dt^k}\psi_X(0). \tag{18}$$

If X is nonnegative and integer-valued then we define its *probability-generating function* $\Phi_X(t)$ to be the power series

$$\Phi_X(t) = \mathbf{E}t^X = \sum_x p_X(k)t^k. \tag{19}$$

Its domain is a symmetric interval about $t = 0$ with radius $r \geq 1$ equal to the radius of convergence for this power series. In order for $r > 1$ it is necessary that X have moments of all orders. These moments are then recoverable through

$$\mathbf{E}X(X-1)\ldots(X-k+1)=\frac{d^k}{dt^k}\Phi_X(1). \tag{20}$$

The distribution of X is always recoverable from Φ_X as

$$p_X(k)=\frac{1}{k!}\frac{d^k}{dt^k}\Phi_X(0).$$

We conclude our discussion of discrete random variables with the following.

Chebyshev's Inequality. *Suppose* X *has a second moment. Then for any* t > 0

$$\mathbb{P}(|X-\mu|\geq t)\leq\frac{\sigma^2}{t^2}, \tag{21}$$

where $\mu = \mathbf{E}\mathrm{X}$ and $\sigma^2 = \mathrm{Var(X)}$.

PROOF. Define

$$f=tI_A,\quad \text{where } A=\{|x-\mu|\geq t\}.$$

Since $|x-\mu|\geq f(x)$ we have

$$\mathbf{E}|X-\mu|^2\geq\mathbf{E}f^2(X)=t^2\mathbb{P}(|X-\mu|\geq t). \qquad \square$$

Chebyshev's inequality is one way of quantifying the fact that σ^2 is a measure of the "spread" of X about its mean. The smaller the value of σ, the more concentrated X is about its mean.

The following is a list of some basic discrete distributions, together with some qualitative and quantitative descriptions.

Basic Distributions (from Parzen [45])

Bernoulli $(0\leq p\leq 1)$

This is a random variable that takes on the values 1 (success) and 0 (failure) with respective probabilities p and $q=1-p$. Trials that can result in either success or failure are called *Bernoulli trials.*

$$p_X(x)=\begin{cases} q, & x=0\\ p, & x=1\end{cases}$$

$$\psi_X(t)=pe^t+q,\qquad \varphi_X(u)=pe^{iu}+q$$

$$\mu=p,\qquad \sigma^2=pq,\qquad \mathbf{E}(X-\mu)^3=pq(q-p),$$

$$\mathbf{E}(X-\mu)^4=3p^2q^2+pq(1-6pq).$$

Binomial $(n = 1, 2, \ldots; 0 \le p \le 1)$

This models the number of successes in n independent Bernoulli trials, in which the probability of success at each trial is p. It also arises in random sampling with replacement.

$$p_X(x) = \binom{n}{x} p^x q^{n-x}; \qquad x = 0, 1, \ldots, n$$

$$\psi_X(t) = (pe^t + q)^n, \qquad \varphi_X(u) = (pe^{iu} + q)^n$$

$$\mu = np, \qquad \sigma^2 = npq, \qquad \mathbf{E}(X - \mu)^3 = npq(q - p),$$

$$\mathbf{E}(X - \mu)^4 = 3n^2p^2q^2 + npq(1 - 6pq).$$

Hypergeometric $(r = 1, 2, \ldots; n, r_1 = 0, 1, \ldots, r)$

This models the number of objects of type one present in a sample of size n drawn from a population of r objects, r_1 being of type one and $r_2 = r - r_1$ being of type two. Sampling is done without replacement (hence the requirement $n \le r$).

$$p_X(x) = \frac{\binom{r_1}{x}\binom{r_2}{n-x}}{\binom{r}{n}}; \qquad \max(0, n - r_2) \le x \le \min(n, r_1)$$

$$\Phi_X(t) = \frac{\binom{r_2}{n}}{\binom{r}{n}} F(\alpha, \beta, \gamma, t)$$

where $\alpha = -n$, $\beta = -r_1$, $\gamma = r_2 - n + 1$, and F is the *hypergeometric function*[1]

$$F(\alpha, \beta, \gamma, t) = 1 + \frac{\alpha\beta}{\gamma}\frac{t}{1!} + \frac{\alpha(\alpha + 1)\beta(\beta + 1)}{\gamma(\gamma + 1)}\frac{t^2}{2!} + \cdots$$

$$\mu = np, \qquad \sigma^2 = npq\left(\frac{r - n}{r - 1}\right),$$

$$\mathbf{E}(X - \mu)^3 = npq(q - p)\left(\frac{r - n}{r - 1}\right)\left(\frac{r - 2n}{r - 2}\right),$$

[1] If $r_2 < n$ then $\gamma \le 0$ and by writing out the factors $\binom{r_2}{n} = \frac{\gamma(\gamma + 1)\ldots r_2}{n!}$ and canceling them with the factors of γ in the denominators of the coefficients for F, one gets a series whose lowest power is $1 - \gamma$.

$$\mathbf{E}(X-\mu)^4 = \frac{npq(r-n)}{(r-1)(r-2)(r-3)}\{r(r+1) - 6n(r-n) + 3pq[r^2(n-2) - rn^2 + 6n(r-n)]\}$$

where $p = r_1/r$ and $q = 1 - p$.

The function F is a polynomial solution to the differential equation

$$t(t-1)\frac{d^2F}{dt^2} + [\gamma - (\alpha + \beta + 1)t]\frac{dF}{dt} - \alpha\beta F = 0.$$

Geometric $(0 \le p \le 1)$

This models the number of trials required to achieve the first success in a sequence of independent Bernoulli trials, in which the probability of success at each trial is p.

$$p_X(x) = pq^{x-1}; \qquad x = 1, 2, \ldots$$

$$\psi_X(t) = \frac{pe^t}{1-qe^t}, \qquad \varphi_X(u) = \frac{pe^{iu}}{1-qe^{iu}}$$

$$\mu = 1/p, \qquad \sigma^2 = q/p^2, \qquad \mathbf{E}(X-\mu)^3 = \frac{q}{p^2}\left(1 + 2\frac{q}{p}\right),$$

$$\mathbf{E}(X-\mu)^4 = \frac{q}{p^2}\left(1 + 9\frac{q}{p^2}\right).$$

Some authors define the geometric distribution in terms of the number of failures encountered until the first success is obtained. This corresponds to replacing X with $X - 1$.

Negative Binomial $(r > 0; 0 \le p \le 1)$

This models the number of failures encountered in a sequence of independent Bernoulli trials (with probability p of success at each trial) before achieving the rth success.

$$p_X(x) = \binom{r+x-1}{x}p^r q^x; \qquad x = 0, 1, \ldots$$

$$\psi_X(t) = \left(\frac{p}{1-qe^t}\right)^r, \qquad \varphi_X(u) = \left(\frac{p}{1-qe^{iu}}\right)^r.$$

$$\mu = rq/p, \qquad \sigma^2 = rq/p^2, \qquad \mathbf{E}(X-\mu)^3 = \frac{rq}{p^2}\left(1 + 2\frac{q}{p}\right),$$

$$\mathbf{E}(X-\mu)^4 = \frac{rq}{p^2}\left[1 + (6+3r)\frac{q}{p^2}\right].$$

We can use the negative binomial distribution even if the parameter r is nonintegral. The coefficient $\binom{r+x-1}{x}$ is evaluated as

$$\frac{r(r+1)\dots(r+x-1)}{x!}.$$

Poisson ($\lambda > 0$)

This models the number of occurrences of events of a specified type in a period of time of length 1, when events of this type are occurring randomly at a mean rate λ per unit time. Many counting time random phenomena are known from experience to be approximately Poisson distributed. Some examples of such phenomena are the number of atoms of a radioactive substance that disintegrate in a unit time interval, the number of calls that come into a telephone exchange in a unit time interval, the number of misprints on a page of a book, and the number of bacterial colonies that grow on a Petri dish that has been smeared with a bacterial suspension.

$$p_X(x) = e^{-\lambda}\frac{\lambda^x}{x!}; \qquad x = 0, 1, \dots$$

$$\psi_X(t) = \exp[\lambda(e^t - 1)], \qquad \varphi_X(u) = \exp[\lambda(e^{iu} - 1)],$$

$$\mu = \lambda, \qquad \sigma^2 = \lambda, \qquad \mathbf{E}(X-\mu)^3 = \lambda, \qquad \mathbf{E}(X-\mu)^4 = \lambda + 3\lambda^2.$$

Absolutely Continuous Random Variables

These are random variables X with distribution satisfying

$$\mathbb{P}(X \in A) = \int_A f_X(x)\, dx, \tag{22}$$

for all Borel subsets $A \subseteq \mathbb{R}$, where f is a nonnegative integrable function defined on $\mathbb{R}$,

$$\int_{-\infty}^{\infty} f_X(x)\, dx = 1. \tag{23}$$

We refer to f_X as the *density* of X. It is sometimes more convenient to work with the *(cumulative) distribution function*

$$F_X(x) = \mathbb{P}(X \leq x) = \int_{-\infty}^{x} f_X(y)\, dy. \tag{24}$$

This function F_X is a continuous nondecreasing function with $F_X(-\infty) = 0$ and $F_X(\infty) = 1$, where we have introduced the notation $F_X(-\infty) =$

$\lim_{x \to -\infty} F_X(x)$ and $F_X(\infty) = \lim_{x \to \infty} F_X(x)$. The connection between f_X and F_X is simply

$$f_X = \frac{dF_X}{dx}. \tag{25}$$

We say that X has *finite expectation* if $\int_{-\infty}^{\infty} |x| f_X(x)\, dx < \infty$. If this holds then we define the *expectation* $\mathbf{E}X$ of X to be

$$\mathbf{E}X = \int_{-\infty}^{\infty} x f_X(x)\, dx. \tag{26}$$

Suppose g is a one-to-one differentiable function and consider the random variable $Y = g(X)$. The distribution function of Y is

$$F_Y(y) = \mathbb{P}(g(X) \leq y) = \begin{cases} F_X(g^{-1}(y)), & \dfrac{dg}{dx} > 0 \\ 1 - F_X(g^{-1}(y)), & \dfrac{dg}{dx} < 0. \end{cases}$$

In any event, we see that Y is absolutely continuous and

$$f_Y(y) = f_X(x) \Big/ \left| \frac{d}{dx} g(x) \right|, \qquad x = g^{-1}(y). \tag{27}$$

If g is not one to one, then one needs to sum over the various inverse branches. Thus, in gereral,

$$f_Y(y) = \sum_{g(x)=y} f_X(x) \Big/ \left| \frac{d}{dx} g(x) \right|. \tag{28}$$

By making the substitution $y = g(x)$ we see that

$$\int |y| f_Y(y)\, dy = \int_{-\infty}^{\infty} |g(x)| f_X(x)\, dx,$$

and thus $g(X)$ has finite expectation if and only if $\int_{-\infty}^{\infty} |g(x)| f_X(x)\, dx < \infty$. In this case, a similar calculation shows that

$$\mathbf{E}g(X) = \int_{-\infty}^{\infty} g(x) f_X(x)\, dx. \tag{29}$$

Suppose that instead of being differentiable, g has only a countable range. Then $Y = g(X)$ is discrete, with

$$p_Y(y) = \int_{g^{-1}(y)} f_X(x)\, dx.$$

In this case

$$\sum_y |y| p_Y(y) = \int_{-\infty}^{\infty} |g(x)| f_X(x)\, dx,$$

and we have the same conclusion as earlier. Namely, $g(X)$ has finite expectation if and only if $\int_{-\infty}^{\infty} |g(x)| f_X(x)\, dx < \infty$, and in this case (29) is valid.

Up to this point we have only considered discrete and absolutely continuous random variables. Thus if X is absolutely continuous then we can only discuss now random variables $Y = g(X)$ for functions g that are either differentiable or at most countably valued.

Many of the results stated earlier for discrete random variables remain true for absolutely continuous random variables; in fact since many of the preceding proofs were formulated in a way not explicitly involving the discrete density function p_X, they carry over as well. In (P_1) we have to limit ourselves to complex-valued functions f_i whose real and imaginary parts are either differentiable or at most countably valued. In (P_2) the equality $\mathbf{E}X = 0$ will never hold since an absolutely continuous random variable X cannot be equal to zero (with probability one). The functions f and g have to be either differentiable or countably valued. Property (P_3) remains valid, and the proofs for properties (P_1)–(P_3) (modified as described earlier) remain valid from the discrete case if one simply replaces sums with integrals. Properties (P_4) and (P_5) need to be restated as follows.

(P_4') *If* X *is (absolutely continuous and) nonnegative real-valued, then* X *has finite expectation if and only if the integral* $\int_0^{\infty} \mathbb{P}(X \geq x)\, dx$ *converges, and moreover this obtains if and only if the series* $\sum_{k=1}^{\infty} \mathbb{P}(X \geq k)$ *converges. If they do converge, then*

$$\sum_{k=1}^{\infty} \mathbb{P}(X \geq k) \leq \mathbf{E}X = \int_0^{\infty} \mathbb{P}(X \geq x)\, dx \leq 1 + \sum_{k=1}^{\infty} \mathbb{P}(X \geq k).$$

(P_5') **(Jensen's Inequality).** *If* X *(is absolutely continuous and) has finite expectation and if* g *is a differentiable convex function defined on* $\mathbb{R}$ *for which* g(X) *has finite expectation then*

$$g(\mathbf{E}X) \leq \mathbf{E}g(X).$$

PROOFS OF (P_4') AND (P_5'). (P_4') Define $Y = [X]$. Then Y is discrete and $Y \leq X \leq Y + 1$. Thus X has finite expectation if and only if Y does. By applying (P_4) to Y we see that this is the case if and only if

$$\sum_{k=1}^{\infty} \mathbb{P}(X \geq k) = \sum_{k=1}^{\infty} \mathbb{P}(Y \geq k) < \infty,$$

in which case $\mathbf{E}Y = \sum_{k=1}^{\infty} \mathbb{P}(X \geq k)$.

Observe now that if X has finite expectation then

$$\lim_{x \to \infty} x[1 - F_X(x)] = 0, \tag{30}$$

since

$$x[1 - F_X(x)] = x \int_x^{\infty} f_X(y)\, dy \leq \int_x^{\infty} y f_X(y)\, dy.$$

Similarly if $\int_0^{\infty} \mathbb{P}(X \geq x)\, dx < \infty$ then X must have finite expectation and (30) must hold, since

$$\int_0^{\infty} \mathbb{P}(X \geq x)\, dx \geq \sum_{k=1}^{\infty} \mathbb{P}(X \geq k).$$

When (30) does hold, it follows upon integrating by parts that

$$\begin{aligned} \int_0^{\infty} x f_X(x)\, dx &= \int_0^{\infty} [1 - F_X(x)]\, dx - x[1 - F_X(x)]\Big/_{x=0}^{\infty} \\ &= \int_0^{\infty} [1 - F_X(x)]\, dx. \end{aligned}$$

□

(P_5') For any x and y we have

$$g(x) - g(y) \geq g'(y)(x - y),$$

since g is convex and differentiable. Thus

$$g(X) - g(\mathbf{E}X) \geq g'(\mathbf{E}X)(X - \mathbf{E}X).$$

Applying expectation to both sides of this inequality leads to the desired conclusion. □

The various statistics (standard deviation, variance, moments, central moments, cumulants) and the characteristic and moment generating functions associated with an absolutely continuous random variable X are defined exactly as we defined them for discrete random variables. Chebyshev's inequality still holds and is proved exactly as earlier. Similarly the properties of characteristic functions still hold. In fact other than (P_5) the proofs given earlier of these properties and Chebyshev's inequality never made explicit use of the discrete density p_X, and so they carry right over to the absolutely continuous setting. Property (P_5), on the other hand, made explicit use of inversion formula (14) for discrete random variables. So this property needs to be addressed now.

When X is absolutely continuous with density f_X, the characteristic function φ_X turns out to be the Fourier transform (or integral) of f_X; namely,

$$\varphi_X(u) = \int_{-\infty}^{\infty} e^{iux} f_X(x)\,dx.$$

It is special in that the L_1-function f_X is nonnegative with total integral one. The distribution of X can be recovered through the inversion formula

$$F_X(x_2) - F_X(x_1) = \frac{1}{2\pi} \lim_{R\to\infty} \int_{-R}^{R} \frac{e^{-iux_1} - e^{-iux_2}}{iu} \varphi_X(u)\,du. \tag{31}$$

In particular if φ_X is integrable, then

$$f_X(x) = \frac{1}{2\pi} \int_{-\infty}^{\infty} e^{-iux} \varphi_X(u)\,du. \tag{32}$$

PROOF. We now prove (31). By interchanging the order of integration (Result II from the earlier analysis)

$$\frac{1}{2\pi} \int_{-R}^{R} \frac{e^{-iux_1} - e^{-iux_2}}{iu} \varphi_X(u)\,du = \mathbf{E} g_R(X) \tag{33}$$

where

$$g_R(y) = \frac{1}{\pi} \int_0^R \left[\frac{\sin u(y - x_1)}{u} - \frac{\sin u(y - x_2)}{u} \right] du.$$

Using the fact that

$$\lim_{R\to\infty} \int_0^R \frac{\sin u}{u}\,du = \pi/2,$$

we see that for $x_1 < x_2$

$$\lim_{R\to\infty} g_R(y) = I_{(x_1, x_2)}(y); \qquad y \neq x_1, x_2.$$

Furthermore since $\displaystyle\int_0^R \frac{\sin u}{u}\,du$ is bounded for all R we see that g_R is bounded. Thus by using Result I from the preceding analysis, we can let $R \to \infty$ in (33) and arrive at (31). □

The following is a list of some basic absolutely continuous distributions, together with some qualitative and quantitative descriptions.

Basic Distributions (from Parzen [45])

Uniform $(a < b)$

This models the location on a line of a dart tossed in such a way that it always hits between the endpoints of the interval a to b, and any two subintervals (of the interval a to b) of equal length have an equal chance of being hit. Similarly if a well-balanced dial is spun around and comes to rest after a large number of revolutions, it is reasonable to assume that the angle of the dial after it stops moving is uniformly distributed on $(0, 2\pi)$. Often in numerical analysis it is assumed that the rounding error caused by dropping all digits more than n places beyond the decimal point is uniformly distributed on $(0, 10^{-n})$.

$$f_X(x) = \frac{1}{b-a}, \qquad a < x < b$$

$$\psi_X(t) = \frac{e^{tb} - e^{ta}}{t(b-a)}, \qquad \varphi_X(u) = \frac{e^{iub} - e^{iua}}{iu(b-a)}$$

$$\mu = \frac{a+b}{2}, \qquad \sigma^2 = \frac{(b-a)^2}{12}, \qquad \mathbf{E}(X-\mu)^3 = 0,$$

$$\mathbf{E}(X-\mu)^4 = \frac{(b-a)^4}{80}.$$

Normal $(-\infty < m < \infty, \sigma > 0)$

Normally distributed random variables occur most often in practical applications. Maxwell's law in physics asserts that under appropriate conditions the components of the velocity of a molecule of gas will be normally distributed, with σ^2 determined from certain physical quantities. Many random variables of interest have distributions that are approximately normal. Thus measurement errors in physical experiments, variability of outputs from industrial production lines, and biological variability (e.g., height and weight) have been found empirically to have approximately normal distributions. It has also been found, both empirically and theoretically, that random fluctuations that result from a combination of many unrelated causes, each individually insignificant, tend to be approximately normally distributed. Theoretical results in this direction are known as "central limit theorems." The number of successes in n independent Bernoulli trials, n large (probability of success p at each trial), approximately obeys a normal probability law with $m = np$, $\sigma^2 = npq$.

$$f_X(x) = \frac{1}{\sigma\sqrt{2\pi}} \exp\left[-\frac{1}{2}\left(\frac{x-m}{\sigma}\right)^2\right], \qquad -\infty < x < \infty$$

$$\psi_X(t) = \exp\left(tm + \frac{1}{2}t^2\sigma^2\right), \qquad \varphi_X(u) = \exp\left(ium - \frac{1}{2}u^2\sigma^2\right)$$

$$\mu = m, \qquad \sigma^2 = \sigma^2, \qquad \mathbf{E}(X-\mu)^3 = 0, \qquad \mathbf{E}(X-\mu)^4 = 3\sigma^4.$$

Exponential ($\lambda > 0$)

The exponential distribution models decay times for radioactive particles. It also models the waiting time required to observe the first occurrence of an event of a specified type when events of this type are occurring randomly at a mean rate λ per unit time. Examples of such waiting times are the time until a piece of equipment fails, the time it takes to complete a job, or the time it takes to get a new customer. It serves as the waiting time distribution for the Poisson counting random variable.

$$f_X(x) = \lambda e^{-\lambda x}, \qquad x > 0$$

$$\psi_X(t) = \frac{\lambda}{\lambda - t}, \qquad \varphi_X(u) = \left(1 - \frac{iu}{\lambda}\right)^{-1}$$

$$\mu = 1/\lambda, \qquad \sigma^2 = 1/\lambda^2, \qquad \mathbf{E}(X-\mu)^3 = 2/\lambda^3, \qquad \mathbf{E}(X-\mu)^4 = 9/\lambda^4$$

Gamma ($r > 0, \lambda > 0$)

This models the waiting time required to observe the rth occurrence of an event of a specified type when events of this type are occurring randomly at a mean rate λ per unit time. There are many applied situations when the density of a random variable can be approximated reasonably well by a gamma density with appropriate parameters.

$$f_X(x) = \frac{\lambda}{\Gamma(r)}(\lambda x)^{r-1}e^{-\lambda x}, \qquad x > 0$$

$$\psi_X(t) = \left(\frac{\lambda}{\lambda - t}\right)^r, \qquad \varphi_X(u) = \left(1 - \frac{iu}{\lambda}\right)^{-r}$$

$$\mu = r/\lambda, \qquad \sigma^2 = r/\lambda^2, \qquad \mathbf{E}(X-\mu)^3 = 2r/\lambda^3,$$

$$\mathbf{E}(X-\mu)^4 = \frac{6r + 3r^2}{\lambda^4}.$$

Chi-Square $(n > 0)$

This models the sum $X_1^2 + \cdots + X_n^2$ of the squares of n independent random variables, each $N(0, 1)$. It corresponds to GAMMA $(r = n/2, \lambda = 1/2)$.

$$f_X(x) = \frac{1}{2^{n/2}\Gamma(n/2)} x^{(n/2)-1} e^{-x/2}, \qquad x > 0$$

$$\psi_X(t) = \left(\frac{1}{1-2t}\right)^{n/2}, \qquad \varphi_X(u) = (1 - 2iu)^{-n/2}$$

$$\mu = n, \qquad \sigma^2 = 2n, \qquad \mathbf{E}(X-\mu)^3 = 8n, \qquad \mathbf{E}(X-\mu)^4 = 12n(n+4).$$

F-Distribution

This models the ratio $\dfrac{U/m}{V/n}$, where U and V are independent random variables, χ^2 distributed with m and n degrees of freedom, respectively.

$$f_X(x) = \frac{\Gamma\left(\frac{m+n}{2}\right)}{\Gamma\left(\frac{m}{2}\right)\Gamma\left(\frac{n}{2}\right)} \left(\frac{m}{n}\right)^{m/2} \frac{x^{(m/2)-1}}{\left(1 + \frac{m}{n}x\right)^{(m+n)/2}}, \qquad x > 0$$

$$\mu = \frac{n}{n-2} \quad (n > 2), \qquad \sigma^2 = \frac{2n^2(m+n-2)}{m(n-2)^2(n-4)} \quad (n > 4)$$

Only the moments of order up to $\left[\dfrac{n-1}{2}\right]$ exist. This distribution is also called the variance ratio distribution and is widely used in statistics for the analysis of variance (ANOVA). It is named after the statistician Sir Ronald Fisher. Related to the F-distribution is the z-distribution, corresponding to the random variable $Z = \frac{1}{2}\log X$.

$$f_Z(z) = \frac{2\Gamma\left(\frac{m+n}{2}\right)}{\Gamma\left(\frac{m}{2}\right)\Gamma\left(\frac{n}{2}\right)} \left(\frac{m}{n}\right)^{m/2} \frac{e^{mz}}{\left(1 + \frac{m}{n}e^{2z}\right)^{(m+n)/2}}, \qquad -\infty < z < \infty$$

$$\varphi_Z(u) = \left(\frac{m}{n}\right)^{u/2} \frac{\Gamma\left(\frac{u+m}{2}\right)\Gamma\left(\frac{u+n}{2}\right)}{\Gamma\left(\frac{m}{2}\right)\Gamma\left(\frac{n}{2}\right)},$$

$$\mu = \frac{1}{2}\left(\frac{1}{n} - \frac{1}{m}\right), \qquad \sigma^2 = \frac{1}{2}\left(\frac{1}{n} + \frac{1}{m}\right).$$

t-Distribution ($n > 0$)

This models the ratio $\dfrac{X}{\sqrt{U/n}}$ where X and U are independent random variables with $N(0, 1)$ and $\chi^2(n)$ distributions, respectively.

$$f_X(x) = \frac{\Gamma\left(\frac{n+1}{2}\right)}{\sqrt{n\pi}\,\Gamma\left(\frac{n}{2}\right)} \frac{1}{\left(1 + \frac{x^2}{n}\right)^{(n+1)/2}}, \qquad -\infty < x < \infty$$

$$\mu = 0, \qquad \sigma^2 = \frac{n}{n-2} \quad (n > 2), \qquad \mathbf{E}(X - \mu)^3 = 0,$$

$$\mathbf{E}(X - \mu)^4 = \frac{3n^2}{(n-2)(n-4)} \quad (n > 4)$$

Only the moments of order up to $n - 1$ exist. This distribution is also called Student's distribution. It is named after "Student" (W.S. Gosset).

Beta ($p > 0, q > 0$)

$$f_X(x) = \frac{1}{B(p, q)} x^{p-1}(1 - x)^{q-1}, \qquad 0 \le x \le 1$$

$$\mu = \frac{p}{p+q}, \qquad \sigma^2 = \frac{pq}{(p+q+1)(p+q)^2}.$$

Cauchy

$$f_X(x) = \frac{1}{\pi(1 + x^2)}, \qquad -\infty < x < \infty$$

$$\varphi_X(u) = e^{-|u|}.$$

None of the moments exist. This corresponds to the t-distribution, $n = 1$.

Rayleigh ($\sigma > 0$)

This models $\sqrt{X^2 + Y^2}$ where X and Y are independent random variables, each $N(0, \sigma^2)$.

$$f_X(x) = \frac{x}{\sigma^2} \exp\left[-\frac{1}{2}\left(\frac{x}{\sigma}\right)^2\right], \qquad x > 0.$$

Distribution Functions

A *distribution function* (abbreviated d.f.) is a real-valued function F defined on $\mathbb{R}$ that is increasing and right continuous with $F(-\infty) = 0$ and $F(\infty) = 1$. If X is any real-valued random variable then F_X defined by

$$F_X(x) = \mathbb{P}(X \leq x)$$

is a d.f., and conversely it can be shown that any d.f. corresponds to a random variable in this way. We define the *point mass* at t, δ_t, to be the d.f. $\delta_t = I_{[t,\infty)}$. When X is a discrete random variable with discrete density p_X,

$$F_X(x) = \sum_y p_X(y)\delta_y(x); \tag{34}$$

and when X is absolutely continuous with density f_X,

$$F_X(x) = \int_{-\infty}^{x} f_X(y)\,dy. \tag{35}$$

Correspondingly we say that a d.f. of the form (34) is *discrete*, and one of the form (35) is *absolutely continuous.*

Since $\sum_x [F(x) - F(x-)] \leq 1$ it follows that a d.f. F can have at most a countably infinite number of jumps. Define the *discrete part* of F to be

$$F_d(x) = \sum_y [F(y) - F(y-)]\delta_y(x). \tag{36}$$

Observe that for $-\infty \leq x_1 < x_2 \leq \infty$

$$F_d(x_2) - F_d(x_1) = \sum_{x_1 < y \leq x_2} [F(y) - F(y-)] \leq F(x_2) - F(x_1) \tag{37}$$

and

$$F_d(x) - F_d(x-) = F(x) - F(x-); \tag{38}$$

this last fact follows by applying a dominated convergence argument (Result I from the preceding analysis) to evaluate the limit

$$\lim_{\varepsilon \downarrow 0} \sum_y [F(y) - F(y-)][\delta_y(x) - \delta_y(x - \varepsilon)],$$

since the jumps $F(y) - F(y-)$ are summable. It follows from (37) that F_d is increasing and inherits the right continuity of F. It also follows by setting $x_1 = -\infty$ and letting $x_2 \to -\infty$ that $F_d(-\infty) = 0$. However, F_d may not be a d.f. since in general $F_d(\infty) \leq 1$.

Consider next the function $F_c = F - F_d$. It is clear that $F_c(-\infty) = 0$ since F and F_d have this property; inequality (37) shows that F_c is increasing. Furthermore F_c is right continuous, being the difference of right continuous functions. More significant is the fact that F_c is also *left* continuous (making it continuous altogether), which follows directly from (38). We refer to F_c as the *continuous part* of F. Again, F_c may not be a d.f. simply because $F_c(\infty) \leq 1$. On the other hand $F_d(\infty) + F_c(\infty) = 1$, so that F can be written as a convex combination of a discrete and continuous d.f. This decomposition is unique since the difference of two discrete d.f.s can be continuous if and only if they are identical.

It is shown in real analysis that if F is a d.f. then the derivative F' exists as an element of L^1. (It exists almost everywhere with respect to Lebesgue measure on $\mathbb{R}$.) Furthermore for $-\infty \leq x_1 < x_2 \leq \infty$

$$\int_{x_1}^{x_2} F'(y)\,dy \leq F(x_2) - F(x_1). \tag{39}$$

We say that F is *singular* if $F' = 0$ almost everywhere. Define the *absolutely continuous part* of F to be

$$F_{ac}(x) = \int_{-\infty}^{x} F'(y)\,dy. \tag{40}$$

Then it follows from (39) that F_{ac} satisfies the same inequality (37) as F_d; namely,

$$F_{ac}(x_2) - F_{ac}(x_1) \leq F(x_2) - F(x_1). \tag{41}$$

So, as we argued for F_c, the function $F_s = F - F_{ac}$ is increasing and right continuous with $F_s(-\infty) = 0$. Furthermore, almost everywhere $F_s' = 0$, so that F_s is singular. We refer to F_s as the *singular part* of F. The discrete parts of F and F_s coincide since F and F_s have the same jumps everywhere. The remainder $F_s - F_d$ must therefore be *singular continuous.* We thus conclude that:

Every d.f. can be written uniquely as a convex combination of a discrete, a singular continuous and an absolutely continuous d.f.

The uniqueness follows from the fact that the difference of two absolutely continuous d.f.s can be singular if and only if they are identical.

A classic example of a singular continuous d.f. is the *Cantor function.* It is defined as follows. If k is the natural number with binary expansion

$$k = \sum_j \varepsilon_j 2^j \qquad (\varepsilon_j = 0 \text{ or } 1)$$

denote

$$k^* = \sum_j \varepsilon_j 3^j.$$

Then the Cantor function is defined on $[0, 1]$ by

$$F(x) = \frac{2k+1}{2^m} \qquad \text{for } x \in \left[\frac{6k^*+1}{3^m}, \frac{6k^*+2}{3^m}\right]$$

$$(m = 1, 2, \ldots \text{ and } k = 0, 1, \ldots, 2^{m-1} - 1).$$

Its continuity follows from the fact that for $x < y$

$$y - x \leq 3^{-m} \Rightarrow F(y) - F(x) \leq 2^{-m}.$$

It arises in the following fashion. Define a sequence of numbers $\{x_n\}$ by $x_1 = 0$ and

$$x_{n+1} = \begin{cases} \frac{1}{3}x_n, & \text{with probability } 1/2 \\ \frac{1}{3}x_n + \frac{2}{3}, & \text{with probability } 1/2; \end{cases}$$

where these probabilities are figured independently at each n. Then with probability one, for all x

$$\lim_{n\to\infty} \frac{\#\text{ of points } x_1, x_2, \ldots, x_n \text{ that are } \leq x}{n} = F(x).$$

In fact what one shows is that this limit is a d.f. satisfying the difference equation

$$2F(x) = F(3x) + F(3x - 2)$$

with boundary conditions

$$F(0) = 0, \qquad F(1) = 1.$$

It can easily be seen that the unique d.f. satisfying these requirements is the Cantor function. (Show directly that $F(x) = 1/2$ for $x \in [1/3, 2/3]$, $F(x) = 1/4$ for $x \in [1/9, 2/9]$, $F(x) = 3/4$ for $x \in [7/9, 8/9]$, $F(x) = 1/8$ for $x \in [1/27, 2/27]$, etc.).

Given a d.f. F and a real-valued function g defined on $\mathbb{R}$, we construct the Riemann-Stieltjes integral as follows. Let $\Pi = (x_0, x_1, \ldots, x_n)$, where

$$a = x_0 < x_1 < \cdots < x_n = b,$$

be any partition of $[a, b]$ and denote

$$\rho(\Pi) = \max_{1\leq i\leq n} (x_i - x_{i-1}).$$

Let $x_i' \in [x_{i-1}, x_i]$, $1 \leq i \leq n$. If

$$\lim_{\rho(\pi)\to 0} \sum_{i=1}^{n} g(x_i')[F(x_i) - F(x_{i-1})]$$

exists then we call it the *Riemann-Stieltjes integral of* g *over* [a, b], denoted

$$\int_a^b g(x)\, dF(x).$$

If this exists for all $a < b$, and if

$$\lim_{\substack{b\to\infty \\ a\to-\infty}} \int_a^b g(x)\, dF(x)$$

also exists, then we call it the *Riemann-Stieltjes integral of* g (*over* $\mathbb{R}$), denoted

$$\int g(x)\, dF(x).$$

When F is discrete and of the form (34), then this integral corresponds to

$$\sum_x g(x) p_X(x);$$

when F is absolutely continuous and of the form (35), then this integral corresponds to

$$\int g(x) f_X(x)\, dx.$$

When g is complex-valued we define the Riemann-Stieltjes integral through the real and imaginary parts of g.

Let X be a random variable with d.f. F_X. Then for any Borel subset $A \subseteq \mathbb{R}$

$$\mathbb{P}(X \in A) = \int I_A(x)\, dF(x) = \int_A dF(x).$$

We say that X has *finite expectation* if $\int |x|\, dF(x) < \infty$, in which case

$$\mathbf{E}X = \int x\, dF(x).$$

The properties of expectation stated for discrete and absolutely continuous random variables are valid in general as well. The *characteristic function* φ_X is defined by

$$\varphi_X(u) = \int e^{iux}\, dF(x).$$

The properties stated earlier for characteristic functions also remain valid. The inversion formula (31) holds whenever x_1 and x_2 are points of continuity of F_X, and this uniquely determines F_X. (The proof involves a more delicate analysis of (33).) The inversion formula (14) also holds, where $p_X(x)$ is interpreted as the jump $F(x) - F(x-) = \mathbb{P}(X = x)$.

Computer Generation of Random Variables

First, suppose that our computer can generate a random variable U that is uniformly distributed on (0, 1). We are given a d.f. F and desire to generate a random variable X having F as its d.f. Define

$$F^{-1}(u) = \min(x \in \mathbb{R} : F(x) \geq u), \qquad u \in (0, 1). \tag{42}$$

Since F is right continuous this minimum exists. Observe that according to this definition

$$F^{-1}(u) \le x \Leftrightarrow u \le F(x).$$

Thus

$$\mathbb{P}(F^{-1}(U) \le x) = \mathbb{P}(U \le F(x)) = F(x),$$

and we see that $X = F^{-1}(U)$ has the desired d.f. If F has the discrete form $\sum p_i \delta_{a_i}$ where $a_1 < a_2 < \cdots$, then (42) leads to

$$F^{-1}(u) = a_i \qquad \text{for } \sum_{j<i} p_j < u \le \sum_{j \le i} p_j.$$

One method of generating U is the *multiplicative congruential method.* The samples are generated according to the recursion

$$x_i = c x_{i-1} \bmod (2^{31} - 1).$$

Each x_i is then scaled into $(0, 1)$. If the multiplier c is a primitive root modulo $2^{31} - 1$ (which is prime), then the generator will have the maximum period of $2^{31} - 2$. Some popular values for c are 16,807 or 397,204,094 or 950,706,376. Typically one can set the seed (if so desired) to any number between 1 and $2^{31} - 2$.

Another method is to use a sequence like

$$x_i = c x_{i-1} + d \qquad (\text{mod } c - 1)$$

where $c = 2^{17}$ and $d = 2^{13} - 1$. Each x_i is then scaled into $[0, 1)$. The period here will be $c - 1$, since it is prime.

Exercises

1. (Parzen [45]) It is estimated that the probability of detecting a moderate attack of tuberculosis using an x-ray photograph of the chest is 0.6. In a city with 60,000 inhabitants, a mass x-ray survey is planned so as to detect all the people with tuberculosis. Two x-ray photographs will be taken of each individual, and he or she will be judged a suspect if at least one of these photographs is found to be "positive." Suppose that in the city there are 2000 persons with moderate attacks of tuberculosis. Let X denote the number of them who, as a result of the survey, will be judged "suspects." Find the mean and variance of X.

2. (Parzen [45]) A man with n keys wants to open his door. He tries the keys independently and at random. Let N_n be the number of trials required to open the door. Find $\mathbf{E}N_n$ and $\text{Var}(N_n)$ if

   (i) unsuccessful keys are not eliminated from further selections.
   (ii) they are.

   Assume that exactly one of the keys can open the door.

3. (Parzen [45]) Let U be a random variable uniformly distributed on the interval $[0, 1]$. Let $g(u)$ be a nondecreasing function, defined for $0 \le u \le 1$.

Find $g(u)$ if the random variable $g(U)$ has the given distribution.

(a) $g(U)$ is Cauchy.
(b) $g(U)$ is exponential, mean $1/\mu$.
(c) $g(U)$ is $N(m, \sigma^2)$.

4. (Parzen [45]) Find the mean and variance of $\cos \pi X$, where X has the given distribution.

(a) X is $N(m, \sigma^2)$.
(b) X is Poisson, mean λ.
(c) X is uniformly distributed on $[-1, 1]$.

5. (Gnedenko [23]) Find $\mathbf{E}X$ and $\mathrm{Var}(X)$ for each of the following random variables.

(a) (Pascal distribution)

$$\mathbb{P}(X = k) = \frac{a^k}{(1 + a)^{k+1}}; \qquad k = 0, 1, 2, \ldots$$

where $a > 0$.

(b) (Polya distribution)

$$\mathbb{P}(X = k) = \left(\frac{a}{1 + \alpha\beta}\right)^k \frac{(1 + \beta)(1 + 2\beta)\ldots(1 + (k-1)\beta)}{k!} p_0;$$

$$k = 0, 1, 2, \ldots$$

where

$$p_0 = \mathbb{P}(X = 0) = (1 + \alpha\beta)^{-1/\beta}.$$

(c) (Laplace distribution)

$$f_X(x) = \frac{1}{2\alpha} e^{-(|x-a|)/\alpha}.$$

(d) (Lognormal distribution)

$$f_X(x) = \frac{1}{\beta x \sqrt{2\pi}} \exp\left[-\frac{(\log x - \alpha)^2}{2\beta^2}\right], \qquad x > 0.$$

6. (Gnedenko [23])

(a) The density function of the magnitude of the velocity of a molecule is given by the Maxwell distribution

$$f_X(x) = \frac{4x^2}{\alpha^3\sqrt{\pi}} e^{-x^2/\alpha^2}, \qquad x > 0,$$

where $\alpha > 0$. Find the average speed and the average kinetic energy of a molecule (the mass of a molecule is m), and the variances of the speed and kinetic energy.

(b) The probability density of the distance x from a reflecting wall at which a molecule in Brownian motion will be found at time $t_0 + t$ if it was at a distance of x_0 from the wall at time t_0 is given by the expression

$$f(x) = \frac{1}{2\sqrt{\pi D t}} \left\{\exp\left[-\frac{(x + x_0)^2}{4Dt}\right] + \exp\left[-\frac{(x - x_0)^2}{4Dt}\right]\right\}, \qquad x \geq 0.$$

Find the expectation and variance of the magnitude of the displacement of the molecule during the time from t_0 to $t_0 + t$.

7. (Gnedenko [23])
   (a) A random variable X is normally distributed. Find $\mathbf{E}|X - m|$, where $m = \mathbf{E}X$.
   (b) Let X be the number of occurences of an event A in n independent trials in each of which $\mathbb{P}(A) = p$. Find $\mathbf{E}X^3$, $\mathbf{E}X^4$, and $\mathbf{E}|X - np|$.

8. (Gnedenko [23])
   (a) Find the characteristic function corresponding to each of the probability density functions.

$$f(x) = \frac{a}{2} e^{-a|x|}.$$

$$f(x) = \frac{a}{\pi(a^2 + x^2)}.$$

$$f(x) = \frac{a - |x|}{a^2}, \qquad |x| \leq a.$$

$$f(x) = \frac{2 \sin^2 \dfrac{ax}{2}}{\pi a x^2}.$$

   (b) Find the probability distribution of each of the random variables whose characteristic function is equal to

$$\varphi(u) = \cos u.$$

$$\varphi(u) = \cos^2 u.$$

$$\varphi(u) = \frac{a}{a + iu}.$$

$$\varphi(u) = \frac{\sin au}{au}.$$

Section II

Multivariate Random Variables

Joint Random Variables

Until now we have been restricted in our consideration of two random variables X and Y, together. We could only talk about, say, the distribution of $X + Y$ or some function $f(X, Y)$, in the special case where Y is a (Borel) function of X, or where X and Y are both (Borel) functions of some third random variable Z. Now we shall dicuss the analysis of joint random variables X and Y in a more general setting.

We begin again with the discrete case. We say that X and Y are *joint discrete random variables* if

$$\mathbb{P}(X = x, Y = y) = p_{XY}(x, y),$$

where p_{XY} satisfies $p_{XY} \geq 0$ and $\sum_x \sum_y p_{XY}(x, y) = 1$. This function p_{XY} is referred to as the *joint discrete density* of X and Y. Whenever X and Y are joint discrete it follows that each of X and Y is discrete, and we can recover their individual densities, called *marginals*, as follows:

$$\mathbb{P}(X = x) = p_X(x) = \sum_y p_{XY}(x, y)$$

$$\mathbb{P}(Y = y) = p_Y(y) = \sum_x p_{XY}(x, y).$$

Based on the familiar condition for independence of events A and B,

$$\mathbb{P}(A \cap B) = \mathbb{P}(A)\mathbb{P}(B);$$

we say that the random variables X and Y are *independent* if

$$p_{XY}(x, y) = p_X(x)p_Y(y). \tag{1}$$

Aside from the individual statistics of X and Y alone, there are several important joint statistics. In general if $Z = f(X, Y)$, where f is a real-valued function defined on $\mathbb{R}^2$, then Z is also discrete and

$$p_Z(z) = \sum_{f(x,y)=z} p_{XY}(x, y). \tag{2}$$

In particular the sum $Z = X + Y$ has density function given by

$$p_Z(z) = \sum_x p_{XY}(x, z - x).$$

In the special case where X and Y are independent, so that (1) holds, this reduces to the *convolution* formula

$$p_Z(z) = \sum_x p_X(x)p_Y(z - x).$$

It follows from (2) that $Z = f(X, Y)$ has finite expectation if and only if $\sum_x \sum_y |f(x, y)|p_{XY}(x, y) < \infty$; in which case

$$\mathbf{E}f(X, Y) = \sum_x \sum_y f(x, y)p_{XY}(x, y). \tag{3}$$

In particular whenever X and Y have finite expectation, then so does $aX + bY$, for any constants a and b; and

$$\mathbf{E}(aX + bY) = a\mathbf{E}X + b\mathbf{E}Y. \tag{4}$$

This generalizes Property (P_1) for expectation from Section I to allow for joint discrete random variables X and Y. In effect, (4) is stating that the operation of taking expectation is a linear one, from the space of random variables to the space of real numbers (a linear functional). If X and Y are independent and if f factors as $f(x, y) = g(x)h(y)$, then using (1)

$$\sum_x \sum_y |f(x, y)|p_{XY}(x, y) = \left[\sum_x |g(x)|p_X(x)\right]\left[\sum_y |h(y)|p_Y(y)\right].$$

Thus $f(X, Y)$ has finite expectation whenever $g(X)$ and $h(Y)$ both do, and in this case a similar summation shows that

$$\mathbf{E}g(X)h(Y) = \mathbf{E}g(X)\mathbf{E}h(Y). \tag{5}$$

("The expectation of a product is the product of the expectations.")

Suppose X and Y each have second moments. Since $|xy| \le \frac{1}{2}(x^2 + y^2)$ it follows that the product XY has finite expectation. We define the *covariance* of X and Y to be

$$\mathrm{Cov}(X, Y) = \mathbf{E}(X - \mu_X)(Y - \mu_Y),$$

where $\mu_X = \mathbf{E}X$ and $\mu_Y = \mathbf{E}Y$. Observe that on account of the linearity (4) (which extends to any finite number of random variables), the covariance also has the alternate form

$$\mathrm{Cov}(X, Y) = \mathbf{E}XY - \mu_X\mu_Y.$$

We say that X and Y are *uncorrelated* if $\text{Cov}(X, Y) = 0$. In particular it follows from (5) that X and Y are uncorrelated whenever they are independent, but the converse need not hold.

Cauchy-Schwarz Inequality. *If* X *and* Y *have second moments then*

$$(\mathbf{E}XY)^2 \leq \mathbf{E}X^2\mathbf{E}Y^2,$$

with equality if and only if either $Y = cX$ *for some constant* c *or* $X = 0$.

PROOF. If $X = 0$ or $Y = 0$ the result is clear. Otherwise set $t = \sqrt[4]{\mathbf{E}Y^2/\mathbf{E}X^2}$ and observe that

$$0 \leq \mathbf{E}(tX - t^{-1}Y)^2 = 2(\sqrt{\mathbf{E}X^2\mathbf{E}Y^2} - \mathbf{E}XY),$$

with equality if and only if $Y = t^2X$. (See Property (P_2) for expectation from Section I. □

One immediate consequence of this lemma is that

$$|\text{Cov}(X, Y)| \leq \sigma_X\sigma_Y.$$

The ratio

$$\rho_{XY} = \frac{\text{Cov}(X, Y)}{\sigma_X\sigma_Y}$$

is called the *correlation coefficient*. It satisfies $|\rho_{XY}| \leq 1$, with equality if and only if X and Y are linearly dependent.

Formula (3) extends to complex-valued functions f, by considering the real and imaginary parts. The special choice $f(x, y) = \exp[i(ux + vy)]$ leads to the *joint characteristic function*

$$\varphi_{XY}(u, v) = \mathbf{E} \exp[i(uX + vY)].$$

When X and Y are independent it follows from (5) that φ_{XY} factors as

$$\varphi_{XY}(u, v) = \varphi_X(u)\varphi_Y(v).$$

A most important operation with joint random variables is that of *conditioning*. Based on the familiar formula for events,

$$\mathbb{P}(A|B) = \frac{\mathbb{P}(A \cap B)}{\mathbb{P}(B)} \qquad \text{if } \mathbb{P}(B) > 0;$$

we can evaluate the *conditional probability density*

$$\begin{aligned} p_{Y|X}(y|x) &= \mathbb{P}(Y = y|X = x) \\ &= \frac{\mathbb{P}(X = x, Y = y)}{\mathbb{P}(X = x)} = \frac{p_{XY}(x, y)}{p_X(x)} \qquad \text{if } p_X(x) > 0. \end{aligned}$$

When X and Y are independent $p_{Y|X}(y|x) = p_Y(y)$, for any x with $p_X(x) > 0$.

In general we can think of $p_{Y|X}$ as a discrete density in its own right involving y, for fixed x, since it sums to one (over y). Thus we can take expectation relative to $p_{Y|X}$. If $p_X(x) > 0$ and if Y has finite expectation then we define the *conditional expectation*

$$\mathbf{E}(Y|X = x) = \sum_y y p_{Y|X}(y|x).$$

Along the same lines if $p_X(x) > 0$ and if Y has a second moment then we define the *conditional variance*

$$\operatorname{Var}(Y|X = x) = \sum_y y^2 p_{Y|X}(y|x) - [\mathbf{E}(Y|X = x)]^2.$$

The *conditional characteristic function* is defined by

$$\varphi_{Y|X}(u|x) = \sum_y e^{iuy} p_{Y|X}(y|x) \qquad \text{if } p_X(x) > 0.$$

We move along now to examine these same ideas for the case of jointly continuous random variables. We say that X and Y are *joint absolutely continuous* if

$$\mathbb{P}((X, Y) \in A) = \iint_A f_{XY}(x, y)\, dx\, dy$$

for all Borel subsets $A \subseteq \mathbb{R}^2$. The function f_{XY} satisfies $f_{XY} \geq 0$ and

$$\int_{-\infty}^{\infty} \int_{-\infty}^{\infty} f_{XY}(x, y)\, dx\, dy = 1.$$

This function f_{XY} is referred to as the *joint density* of X and Y. Whenever X and Y are joint absolutely continuous it follows that each of X and Y is absolutely continuous, and we can recover their individual densities, or marginals, through

$$\begin{aligned} f_X(x) &= \int_{-\infty}^{\infty} f_{XY}(x, y)\, dy \\ f_Y(y) &= \int_{-\infty}^{\infty} f_{XY}(x, y)\, dx. \end{aligned} \tag{6}$$

When X and Y are *independent* then f_{XY} factors as $f_{XY}(x, y) = f_X(x) f_Y(y)$. Sometimes it is more convenient to work with the joint cumulative distribution function

$$F_{XY}(x, y) = \mathbb{P}(X \leq x, Y \leq y) = \int_{-\infty}^{x} \int_{-\infty}^{y} f_{XY}(x', y')\, dx'\, dy'.$$

The marginal cumulatives are given in terms of F_{XY} by

$$\begin{aligned} F_X(x) &= \mathbb{P}(X \leq x) = F_{XY}(x, \infty) = \lim_{y \to \infty} F_{XY}(x, y) \\ F_Y(y) &= \mathbb{P}(Y \leq y) = F_{XY}(\infty, y) = \lim_{x \to \infty} F_{XY}(x, y). \end{aligned}$$

The joint density is given by

$$f_{XY} = \frac{\partial^2}{\partial x \, \partial y} F_{XY}.$$

Again, when X and Y are independent then F_{XY} factors as $F_{XY}(x, y) = F_X(x)F_Y(y)$.

As regards the distribution of $g(X, Y)$ for real-valued functions g defined on $\mathbb{R}^2$, it is more convenient here to deal with transformations $u = u(x, y)$, $v = v(x, y)$ from $\mathbb{R}^2$ into $\mathbb{R}^2$. We assume that this transformation is one to one and differentiable. Then $U = u(X, Y)$ and $V = v(X, Y)$ are joint absolutely continuous, and in analogy to (I.27),

$$f_{UV}(u, v) = f_{XY}(x, y) \left| \frac{\partial(x, y)}{\partial(u, v)} \right|. \tag{7}$$

This follows from the identity

$$\iint_A f_{XY}(x, y)\, dx\, dy = \iint_{A'} f_{XY}(x, y) \left| \frac{\partial(x, y)}{\partial(u, v)} \right| du\, dv, \tag{8}$$

for the change of coordinates $(x, y) \mapsto (u, v)$, where A' is the set in (u, v)-space corresponding to A in (x, y)-space. For example, setting $u = x + y$, $v = x$ and combining (7) with (6) leads to

$$f_Z(z) = \int_{-\infty}^{\infty} f_{XY}(x, z - x)\, dx$$

for the density of $Z = X + Y$. As earlier, when X and Y are independent this leads to the *convolution formula*

$$f_Z(z) = f_X * f_Y(z) = \int_{-\infty}^{\infty} f_X(x) f_Y(z - x)\, dx.$$

By substituting $|u| f_{XY}$ for f_{XY} in (8) we find that

$$\iint |u| f_{UV}(u, v)\, du\, dv = \iint |u(x, y)| f_{XY}(x, y)\, dx\, dy,$$

so that $u(X, Y)$ has finite expectation if and only if $\iint |u(x, y)| f_{XY}(x, y)$ $dx\, dy < \infty$, in which case

$$\mathbf{E}u(X, Y) = \iint u(x, y) f_{XY}(x, y)\, dx\, dy.$$

The linearity condition (4) is easily seen to hold here as well; the covariance, correlation coefficient, and joint characteristic function are defined exactly as previously. (Observe that the proof of the Cauchy-Schwarz inequality given earlier made no explicit use of the discrete density p_{XY}, and so carries over to the absolutely continuous setting directly.)

The conditional density $f_{Y|X}$ is given by

$$f_{Y|X}(y|x) = \frac{f_{XY}(x, y)}{f_X(x)} \qquad \text{if } f_X(x) > 0.$$

As earlier, we can regard $f_{Y|X}$ as a density in its own right, for each fixed x, since it integrates to one (over y). In particular we can take expectation relative to $f_{Y|X}$. Thus if $f_X(x) > 0$ and if Y has finite expectation then we define the conditional expectation

$$\mathbf{E}(Y|X = x) = \int y f_{Y|X}(y|x)\, dy.$$

If $f_X(x) > 0$ and if Y has a second moment then we define the conditional variance

$$\mathrm{Var}(Y|X = x) = \int y^2 f_{Y|X}(y|x)\, dy - [\mathbf{E}(Y|X = x)]^2.$$

Finally, if $f_X(x) > 0$, the conditional characteristic function is defined by

$$\varphi_{Y|X}(u|x) = \int e^{iuy} f_{Y|X}(y|x)\, dy.$$

Again, when X and Y are independent $f_{Y|X} = f_Y$ for every value of x, $f_X(x) > 0$.

The following result is a generalization of the Cauchy-Schwarz inequality.

Hölder's Inequality. *Let* X *and* Y *be joint random variables. Suppose that* X *has a finite* pth *order moment and that* Y *has a finite* qth *order moment where* p *and* q *are positive numbers satisfying* $\frac{1}{p} + \frac{1}{q} = 1$. *Then* XY *has a finite first moment, and*

$$\mathbf{E}|XY| \le (\mathbf{E}|X|^p)^{1/p}(\mathbf{E}|Y|^q)^{1/q},$$

with equality if and only if $|Y|^q = c|X|^p$ *a.s. for some constant* c *or* $X = 0$ *a.s.*

Proof. It suffices to establish Young's Inequality, which states that for $x, y \ge 0$

$$xy \le \frac{x^p}{p} + \frac{y^q}{q}$$

with equality if and only if $y = x^{p-1}$. Observe now that $\frac{x^p}{p}$ is the area underneath the curve of $f(t) = t^{p-1}$ from $t = 0$ to $t = x$, and $\frac{y^q}{q}$ is the area underneath $f^{-1}(t) = t^{1/(p-1)} = t^{q-1}$ from $t = 0$ to $t = y$. □

Conditional Expectation

Often we are given two joint random variables X and Y and would like to find the "best fit" $Y \approx g(X)$ in some sense. That is, we would like to find that (Borel) function g for which $g(X)$ is as "close" to Y as possible. In many settings this "best approximation" works out to be the conditional expectation $\mathbf{E}(Y|X)$.

Definition I. Let X and Y be joint random variables and suppose Y has finite expectation. The *conditional expectation* $\mathbf{E}(Y|X)$ is the random variable $g(X)$, where g is defined by

$$g(x) = \mathbf{E}(Y|X = x). \tag{9}$$

Thus the function g, which "best fits" X to Y is defined to be the conditional expectation of Y given that $X = x$, at each argument x. It may not be apparent at this stage just why $\mathbf{E}(Y|X)$ should be the "best approximation" to Y, and the ensuing discussion is intended to illuminate this. We begin with a lemma, which gives an alternate characterization of $\mathbf{E}(Y|X)$, one that is usually taken as a starting point in more theoretical discussions.

Lemma I. E(Y|X) *is the (a.s.) unique random variable of the form* g(X), *for* g *a (Borel) function, satisfying the following condition. For any bounded (Borel) function* h,

$$\mathbf{E}[h(X)\mathbf{E}(Y|X)] = \mathbf{E}[h(X)Y]. \tag{10}$$

Proof. We shall prove (10) for the absolutely continuous case, since the proof for the discrete case can be obtained by converting integrals to sums.

$$\begin{aligned}
\mathbf{E}[h(X)\mathbf{E}(Y|X)] &= \int h(x)g(x)f_X(x)\,dx = \int h(x)\mathbf{E}(Y|X = x)f_X(x)\,dx \\
&= \int h(x)\left[\int y f_{Y|X}(y|x)\,dy\right] f_X(x)\,dx \\
&= \iint h(x)y f_{XY}(x, y)\,dy\,dx = \mathbf{E}[h(X)Y].
\end{aligned}$$

To see that (10) uniquely determines $g(X)$, observe that if $g_1(X)$ and $g_2(X)$ both satisfy (10), then their difference $g(X) = g_1(X) - g_2(X)$ satisfies

$$\mathbf{E}[h(X)g(X)] = 0$$

for all bounded (Borel) functions h. By choosing $h = I_{\{g>\varepsilon\}}$ and $h = I_{\{g<-\varepsilon\}}$ and letting $\varepsilon \downarrow 0$ we conclude from this that $\mathbb{P}(g(X) = 0) = 1$. □

Observe that by setting $h(X) = \operatorname{sgn} \mathbf{E}(Y|X)$ in (10) we see that $\mathbf{E}(Y|X)$ has finite expectation, and that by setting $h = 1$ we arrive at

$$\mathbf{E}[\mathbf{E}(Y|X)] = \mathbf{E}Y.$$

The fact that (10) uniquely determines $\mathbf{E}(Y|X)$ can be used to show that some random variable Z is equal to $\mathbf{E}(Y|X)$. One merely needs to show that $\mathbf{E}[h(X)Z] = \mathbf{E}[h(X)Y]$ for bounded (Borel) functions h, and that Z is a (Borel) function of X.

Properties of Conditional Expectation

(P_1) *$\mathbf{E}(Y|X)$ is linear in Y; so that if Y_i have finite expectation and if a_i are constants, $1 \le i \le n$, then*

$$\mathbf{E}\left(\sum_{i=1}^{n} a_i Y_i | X\right) = \sum_{i=1}^{n} a_i \mathbf{E}(Y_i|X).$$

(P_2) *If Y has finite expectation and $Y \ge 0$ then $\mathbf{E}(Y|X) \ge 0$. In particular if Y_1 and Y_2 have finite expectation and if $Y_1 \le Y_2$, then $\mathbf{E}(Y_1|X) \le \mathbf{E}(Y_2|X)$.*

(P_3) *If Y and $h(X)Y$ have finite expectation, then*

$$\mathbf{E}[h(X)Y|X] = h(X)\mathbf{E}(Y|X).$$

Similarly if $h(X)$ has finite expectation then $\mathbf{E}[h(X)|X] = h(X)$.

(P_4) *If Y has finite expectation and if X and Y are independent then*

$$\mathbf{E}(Y|X) = \mathbf{E}Y.$$

(P_5) *If Y has a moment of order $r \ge 1$ then*

$$|\mathbf{E}(Y|X)|^r \le [\mathbf{E}(|Y||X)]^r \le \mathbf{E}(|Y|^r|X),$$

and thus $\mathbf{E}(Y|X)$ also has a moment of order r.

Proofs. (P_1) This follows directly from (10) and the linearity of expectation, since

$$\mathbf{E}\left\{h(X)\left[\sum_{i=1}^{n} a_i \mathbf{E}(Y_i|X)\right]\right\} = \sum_{i=1}^{n} a_i \mathbf{E}[h(X)\mathbf{E}(Y_i|X)]$$

$$= \sum_{i=1}^{n} a_i \mathbf{E}[h(X)Y_i] = \mathbf{E}\left[h(X)\left(\sum_{i=1}^{n} a_i Y_i\right)\right]. \quad \square$$

(P_2) Set $h(X) = I_{\{\mathbf{E}(Y|X)<0\}}$. Since

$$0 \geq \mathbf{E}[h(X)\mathbf{E}(Y|X)] = \mathbf{E}[h(X)Y] \geq 0,$$

it must be that $\mathbb{P}(\mathbf{E}(Y|X) < 0) = 0$. □

(P_3) This follows directly from (10). □

(P_4) This follows directly from (10) since $h(X)$ and Y are independent. □

(P_5) Since $-|Y| \leq Y \leq |Y|$ the first inequality follows from (P_2). To prove the second inequality assume without loss of generality that $Y \geq 0$. By the Mean-Value theorem $y^r - y_0^r \geq (y - y_0)ry_0^{r-1}$ for any y_0, $y \geq 0$. Thus

$$Y^r - [\mathbf{E}(Y|X)]^r \geq [Y - \mathbf{E}(Y|X)]r[\mathbf{E}(Y|X)]^{r-1}.$$

Now take conditional expectation of both sides with respect to X, and use (P_3). □

When Y has a second moment we define its *conditional variance* with respect to X as

$$\operatorname{Var}(Y|X) = \mathbf{E}\{[Y - \mathbf{E}(Y|X)]^2|X\} = \mathbf{E}(Y^2|X) - [\mathbf{E}(Y|X)]^2.$$

It is readily checked that

$$\mathbf{E} \operatorname{Var}(Y|X) + \operatorname{Var} \mathbf{E}(Y|X) = \operatorname{Var} Y. \tag{11}$$

In analogy to (I.9) we observe that for any (Borel) function h such that $h(X)$ has a second moment

$$\mathbf{E}[Y - h(X)]^2 = \mathbf{E} \operatorname{Var}(Y|X) + \mathbf{E}[h(X) - \mathbf{E}(Y|X)]^2.$$

Thus whenever Y has a second moment, $\mathbf{E}(Y|X)$ is the best approximation to Y in the mean-square sense, over all square integrable random variables $h(X)$. Furthermore the mean square error is $\mathbf{E} \operatorname{Var}(Y|X)$. This is elaborated later, when we interpret $\mathbf{E}(Y|X)$ as an orthogonal projection.

Conditioning is most useful when one is studying *compound random variables* for which some parameters are themselves random variables. For example, let $X_1, X_2, \ldots$ be independent identically distributed nonnegative integer-valued random variables with common generating function Φ_X. Let N be a nonnegative integer-valued random variable, independent of the X_ns, with generating function Φ_N. Consider the sum $S = X_1 + \cdots + X_N$. Then

$$\Phi_S(t) = \mathbf{E}t^S = \mathbf{E}[\mathbf{E}(t^S|N)] = \mathbf{E}\Phi_X^N(t) = \Phi_N(\Phi_X(t)).$$

Similarly

$$\mathbf{E}S = \mathbf{E}[\mathbf{E}(S|N)] = \mathbf{E}(N\mathbf{E}X) = \mathbf{E}N\mathbf{E}X,$$

and by (11)

$$\begin{aligned}\operatorname{Var} S &= \mathbf{E}\operatorname{Var}(S|N) + \operatorname{Var}\mathbf{E}(S|N)\\ &= \mathbf{E}(N\operatorname{Var} X) + \operatorname{Var}(N\mathbf{E}X) = \mathbf{E}N\operatorname{Var} X + (\mathbf{E}X)^2\operatorname{Var} N.\end{aligned}$$

This example will be useful in our discussion of branching chains.

Orthogonal Projections

We describe here an alternative approach to conditioning, useful in the mean square setting.

Theorem II. *Let* C *be a closed convex set in a Hilbert space* $\mathscr{H}$. *For every* $x \in \mathscr{H}$ *there is a unique point* $P_C x \in C$ *such that*

$$\|x - P_C x\| = \inf_{y \in C} \|x - y\| = d(x, C).$$

Proof. Let $d = \inf_{y \in C} \|x - y\|$ and let $x_n \in C$ be such that $\|x - x_n\| \to d$ as $n \to \infty$. Since C is convex $\frac{1}{2}(x_n + x_m) \in C$ and thus $\|\frac{1}{2}(x_n + x_m) - x\| \geq d$. From the Parallelogram Identity it follows that

$$\begin{aligned}\|x_n - x_m\|^2 &= \|(x_n - x) - (x_m - x)\|^2\\ &= 2\|x_n - x\|^2 + 2\|x_m - x\|^2 - 4\left\|\frac{x_n + x_m}{2} - x\right\|^2\\ &\leq 2\|x_n - x\|^2 + 2\|x_m - x\|^2 - 4d^2. \qquad (12)\end{aligned}$$

The right-hand side of (12) tends to zero as $m, n \to \infty$ and thus $\{x_n\}$ is a Cauchy sequence. Since $\mathscr{H}$ is complete and C is closed $x_n \to P_C x \in C$ and $\|x - P_C x\| = d$. Suppose next that for some $z \in C$ there also holds $\|z - x\| = d$. If we substitute in (12) $x_n = z$ and $x_m = P_C x$ we obtain $z = P_C x$, from which follows the uniqueness. □

Theorem III. *Let* S *be a closed subspace of* $\mathscr{H}$ *and let* $y \in S$. *Then* $y = P_S x$ *if and only if* $x - y \in S^\perp$.

Proof. For any $z \in S$, $0 < t \leq 1$

$$\begin{aligned}&\|x - y\|^2 - \|x - [y + t(z - y)]\|^2\\ &\qquad = 2t\operatorname{Re}\langle x - y, y - z\rangle - t^2\|z - y\|^2. \qquad (13)\end{aligned}$$

If $x - y \in S^\perp$ then the right-hand side of (13) is less than or equal to zero, and by choosing $t = 1$ we get $\|x - y\| = \inf_{z \in S} \|x - z\|$. Hence $y = P_S x$. Conversely if $y = P_S x$ then the left-hand side of (13) is less than or equal to zero, and thus for each t, $0 < t \leq 1$,

$$2t\,\mathrm{Re}\langle x-y, y-z\rangle - t^2\|z-y\|^2 \le 0.$$

Upon dividing by t and letting $t \downarrow 0$ we obtain $\mathrm{Re}\langle x-y, y-z\rangle \le 0$, $\forall z \in S$. From this, it follows directly that $x - y \in S^\perp$. □

Corollary IV. *Let* S *be a closed subspace of* $\mathscr{H}$.

(i) *Every vector* $\mathrm{x} \in \mathscr{H}$ *can by represented uniquely as*

$$x = y + z, \qquad y \in S, \qquad z \in S^\perp.$$

In this representation $\mathrm{y} = \mathrm{P_S x}$, $\mathrm{z} = \mathrm{P_{S^\perp} x}$. *Furthermore* $\|\mathrm{x}\|^2 = \|\mathrm{y}\|^2 + \|\mathrm{z}\|^2$.

(ii) $\mathrm{P_S}$ *is a linear operator satisfying*

$$P_S^2 = P_S, \qquad \|P_S\| = 1.$$

When S is a subspace of finite dimension n, we can reduce the problem of finding $P_S x$ to a system of n simultaneous linear equations. Let $b_1, \ldots, b_n$ be a basis for S. Let $x \in \mathscr{H}$, $y = P_S x$. Since $y \in S$ there holds

$$y = \sum_{k=1}^{n} \xi_j b_j.$$

From our theorem, we know that $x - y$ is orthogonal to every $z \in S$, in particular to the vectors $b_1, \ldots, b_n$. Thus

$$\langle x - y, b_i\rangle = 0, \qquad 1 \le i \le n.$$

This system of equations can be written explicitly as

$$\sum_{j=1}^{n} \langle b_j, b_i\rangle \xi_j = \langle x, b_i\rangle, \qquad 1 \le i \le n.$$

For our setting we let $\mathscr{H} = L^2(\mathbb{P})$ and let S be the subspace of all square integrable random variables $h(X)$, h being a Borel function. Then it follows from (10) that

$$\mathbf{E}(Y|X) = P_S Y.$$

More generally if $X_1, \ldots, X_n$, Y are joint random variables we can take S to be the subspace of all square integrable random variables $h(X_1, \ldots, X_n)$, h being a Borel function defined on $\mathbb{R}^n$. In this case the problem of finding $\mathbf{E}(Y|X_1, \ldots, X_n)$ can be viewed as *nonlinear estimation*. We are looking for the best approximation $h(X_1, \ldots, X_n)$ to Y in mean square.

In contrast to this, the problem of *linear estimation* looks for the best approximation

$$a + a_1 X_1 + \cdots + a_n X_n$$

to Y, in mean square. Here the subspace S onto which we are projecting is the span of $1, X_1, \ldots, X_n$. Since it is finite-dimensional, by our earlier remark, the linear estimation problem is then equivalent to a linear algebraic system. Precisely, the optimal $a, a_1, \ldots, a_n$ satisfy

$$\left.\begin{aligned} & a + \sum_{j=1}^{n} a_j \mathbf{E} X_j = \mathbf{E} Y \\ & \sum_{j=1}^{n} a_j \operatorname{Cov}(X_i, X_j) = \operatorname{Cov}(X_i, Y), 1 \leq i \leq n \end{aligned}\right\} \tag{14}$$

What is striking is that for the nonlinear estimation problem, where S is infinite-dimensional, as long as $X_1, \ldots, X_n, Y$ have a joint (discrete or absolutely continuous) distribution, *the computation of* $\mathbf{E}(Y|X_1, \ldots, X_n)$ *can be done directly at once from the joint density.* Simply compute (say, for the continuous case)

$$\begin{aligned} g(x_1, \ldots, x_n) &= \mathbf{E}(Y|X_1 = x_1, \ldots, X_n = x_n) \\ &= \int y f_{Y|X_1, \ldots, X_n}(y|x_1, \ldots, x_n)\, dy, \end{aligned}$$

and then

$$\mathbf{E}(Y|X_1, \ldots, X_n) = g(X_1, \ldots, X_n).$$

Joint Normal Distribution

One says that the random vector $X = (X_1, \ldots, X_n)^t$ is *joint normal* if its joint characteristic function is given by

$$\varphi_X(u) = \exp\left(i\langle u, \mu\rangle - \frac{1}{2}\langle \Gamma u, u\rangle\right),$$

where $u = (u_1, \ldots, u_n)^t$; and $\mu = (\mu_1, \ldots, \mu_n)^t$ and $\Gamma = (\gamma_{ij})_{i,j=1}^n$ are given by $\mu_i = \mathbf{E}X_i$ and $\gamma_{ij} = \operatorname{Cov}(X_i, X_j)$, $1 \leq i, j \leq n$. The matrix Γ is symmetric nonnegative definite, since $\langle \Gamma u, u\rangle = \operatorname{Var}(\langle u, X\rangle)$. It is nonsingular if and only if $X_1, \ldots, X_n$ are linearly independent. In any event we can find a square matrix A such that $\Gamma = AA^t$. Let $Z_1, \ldots, Z_n$ be independent random variables, each distributed $N(0, 1)$. Consider the random vector $Y = AZ$, where $Z = (Z_1, \ldots, Z_n)^t$. Then

$$\varphi_Y(u) = \exp\left(-\frac{1}{2}\|A^t u\|^2\right) = \exp\left(-\frac{1}{2}\langle \Gamma u, u\rangle\right),$$

so that X and $Y + \mu$ have the same joint distribution. When Γ is nonsingular we can apply our change of variable formula (7) to the transformation $x = Az + \mu$ and compute the joint density of X as

$$\begin{aligned} f_X(x) = f_Z(z)/|\det A| &= \frac{1}{(2\pi)^{n/2}\sqrt{\det \Gamma}} e^{-1/2\|z\|^2} \\ &= \frac{1}{(2\pi)^{n/2}\sqrt{\det \Gamma}} \exp\left[-\frac{1}{2}\langle \Gamma^{-1}(x - \mu), x - \mu\rangle\right]. \end{aligned}$$

An equivalent definition of joint normality is that every linear combination $\langle u, X\rangle$ be (univariate) normal. This becomes most useful when performing nonlinear operations on X. For example if X_1, X_2, X_3, X_4 are joint normal with zero means, and if $U_i^2 = X_i^2 - \mathbf{E}(X_i^2)$, $1 \le i \le 4$, then (Parzen [45])

$$\operatorname{Cov}(X_1^2, X_2^2) = 2\operatorname{Cov}^2(X_1, X_2)$$

$$\mathbf{E}(X_1X_2X_3X_4) = \mathbf{E}(X_1X_2)\mathbf{E}(X_3X_4) + \mathbf{E}(X_1X_3)\mathbf{E}(X_2X_4) + \mathbf{E}(X_1X_4)\mathbf{E}(X_2X_3)$$

$$\mathbf{E}(U_1U_2U_3U_4) = 8[\mathbf{E}(X_1X_2X_3X_4)]^2 - 4[\mathbf{E}(X_1X_2)\mathbf{E}(X_3X_4)]^2 - 4[\mathbf{E}(X_1X_3)\mathbf{E}(X_2X_4)]^2 - 4[\mathbf{E}(X_1X_4)\mathbf{E}(X_2X_3)]^2$$

If $X_1, \ldots, X_n$ are joint normal, then the fact that they are uncorrelated implies that they are independent, since Γ then becomes a diagonal matrix. We next consider the conditional expectation $\mathbf{E}(Y|X_1, \ldots, X_n)$, where $X_1, \ldots, X_n, Y$ are all joint normal. Assume without loss of generality that Γ, the covariance matrix for the X_is, is nonsingular. If we choose the coefficients $a, a_1, \ldots, a_n$ according to (14) then $Y' = Y - a - \sum_{j=1}^{n} a_jX_j$ satisfies

$$\operatorname{Cov}(Y', X_i) = 0, \qquad 1 \le i \le n.$$

Thus Y' is independent of $X_1, \ldots, X_n$. Furthermore $\mathbf{E}Y' = 0$, and so for any bounded (Borel) function h defined on $\mathbb{R}^n$

$$\mathbf{E}[h(X_1, \ldots, X_n)Y'] = 0.$$

From this we infer by (10) that

$$\mathbf{E}(Y|X_1, \ldots, X_n) = a + \sum_{j=1}^{n} a_jX_j,$$

so that the best nonlinear estimate of Y turns out in fact to be linear.

Furthermore, by applying the change of variable formula (7) to the transformation $(x_1, \ldots, x_n, y') \to (x_1, \ldots, x_n, y)$ where $y' = y - a - \sum_{j=1}^{n} a_jx_j$, we find that

$$f_{Y|X_1,\ldots,X_n}(y|x_1, \ldots, x_n) = f_{Y'}(y').$$

Thus the conditional distribution of Y given $X_1, \ldots, X_n$ is (univariate) normal, with constant conditional variance $\sigma_{Y'}^2 = \operatorname{Cov}(Y, Y')$.

Multidimensional Distribution Functions

A real-valued function F defined on $\mathbb{R}^m$ is said to be a (*multivariate*) *d.f.* if

(i) it is increasing and right continuous in each of its arguments;

(ii) $\lim\limits_{x_k \to -\infty} F(x_1, \ldots, x_m) = 0$, $1 \le k \le m$, and $F(\infty, \ldots, \infty) = 1$;

(iii) for any $a_k < b_k$, $1 \le k \le m$,

$$\sum_{S \subseteq \{1,\dots,m\}} (-1)^{|S|} F_S[a, b] \ge 0,$$

where $F_S[a, b]$ denotes the value $F(x_1, \dots, x_m)$ when $x_k = a_k$, $k \in S$ and $x_k = b_k$, $k \notin S$, and $|S|$ denotes the cardinality of S. If X is a random vector in $\mathbb{R}^m$ then

$$F_X(x_1, \dots, x_m) = \mathbb{P}(X_1 \le x_1, \dots, X_m \le x_m)$$

is a d.f., and conversely every d.f. arises from some random vector X in this way. In this connection observe that the term $\sum_S (-1)|S|F_S[a, b]$ is equal to $\mathbb{P}(a_1 < X_1 \le b_1, \dots, a_m < X_m \le b_m)$. To see that condition (iii) is not redundant in general (as it is when $m = 1$), consider $F = I_A$ where A is the two-dimensional region $\{x, y \ge 0 \text{ and } x + y \ge 1\}$. This function satisfies (i) and (ii), but

$$F(1, 1) - F\left(1, \frac{1}{2}\right) - F\left(\frac{1}{3}, 1\right) + F\left(\frac{1}{3}, \frac{1}{2}\right) = -1.$$

If there exists a function $f(x_1, \dots, x_m)$ such that

$$F(x_1, \dots, x_m) = \int_{-\infty}^{x_1} \cdots \int_{-\infty}^{x_m} f(z_1, \dots, z_m)\, dz_1 \dots dz_m$$

then f is called the *density function* for the d.f. F. It has the features that $f \ge 0$ and $\int_A f(x)\, dx = \mathbb{P}(X \in A)$ for every Borel subset $A \subseteq \mathbb{R}^m$. In particular if f is continuous at the point $x = (x_1, \dots, x_m)$ then up to infinitesimals of higher order, the probability of X falling in the rectangular parallelepiped $x_k \le X_k \le x_k + dx_k$ is given by

$$\mathbb{P}(x_k \le X_k \le x_k + dx_k, 1 \le k \le m) = f(x_1, \dots, x_m)\, dx_1 \dots dx_m.$$

The marginal d.f.s for the X_ks are given by

$$F_k(x_k) = \lim_{x_1, \dots, x_{k-1}, x_{k+1}, \dots, x_m \to \infty} F(x).$$

When F has a density function f so does each F_k, and their densities f_k are given by

$$f_k(x_k) = \int_{-\infty}^{\infty} \cdots \int_{-\infty}^{\infty} f(x)\, dx_1 \dots dx_{k-1}\, dx_{k+1} \dots dx_m.$$

The random variables X_k are *independent* if and only if

$$F(x) = \prod_{k=1}^{m} F_k(x_k),$$

and when F has a density function f this condition becomes

$$f(x) = \prod_{k=1}^{m} f_k(x_k).$$

We say that X has *finite moments of order r* if $\int_{\mathbb{R}^m} \|x\|^r \, dF_X(x) < \infty$. If this is the case then we define the *joint moments*

$$\mathbf{E}X_1^{r_1} \ldots X_m^{r_m} = \int_{\mathbb{R}^m} x_1^{r_1} \ldots x_m^{r_m} \, dF_X(x),$$

$0 \le r_1 + \cdots + r_m \le r$. When the second moments are finite we define the *covariance* by

$$\operatorname{Cov}(X_k, X_l) = \mathbf{E}X_k X_l - \mathbf{E}X_k \mathbf{E}X_l.$$

Let g be a real-valued function defined on $\mathbb{R}^m$. If $\int_{\mathbb{R}^m} |g(x)| \, dF_X(x) < \infty$ then $g(X_1, \ldots, X_m)$ has finite expectation, and

$$\mathbf{E}g(X_1, \ldots, X_m) = \int_{\mathbb{R}^m} g(x) \, dF_X(x).$$

In particular, expectation is a linear operation and covariance is bilinear. Thus if $X_1, \ldots, X_m$ are joint random variables with finite first moments, then

$$\mathbf{E}\left(\sum_{k=1}^m a_k X_k\right) = \sum_{k=1}^m a_k \mathbf{E}X_k;$$

and if $X_1, \ldots, X_m, Y_1, \ldots, Y_n$ are joint random variables with finite second moments then

$$\operatorname{Cov}\left(\sum_{k=1}^m a_k X_k, \sum_{l=1}^n b_l Y_l\right) = \sum_{k=1}^m \sum_{l=1}^n a_k b_l \operatorname{Cov}(X_k, Y_l). \tag{15}$$

In particular if the X_ks are independent then

$$\operatorname{Var}\left(\sum_{k=1}^m a_k X_k\right) = \sum_{k=1}^m a_k^2 \operatorname{Var} X_k.$$

The $m \times m$ matrix $(\operatorname{Cov}(X_k, X_l))_{k,l=1}^m$ is called the *covariance matrix*. It is always symmetric and nonnegative definite, since by (15)

$$\sum_{k=1}^m \sum_{l=1}^m a_k a_l \operatorname{Cov}(X_k, X_l) = \operatorname{Var}\left(\sum_{k=1}^m a_k X_k\right).$$

The *joint characteristic function* is given by

$$\varphi_X(u) = \mathbf{E}(\exp(i\langle u, X\rangle)) = \int_{\mathbb{R}^m} e^{i\langle u, x\rangle} \, dF_X(x), \qquad u \in \mathbb{R}^m.$$

If X has finite moments of order r then φ_X is r times continuously differentiable at the origin, and

$$\mathbf{E}X_1^{r_1} \ldots X_m^{r_m} = \frac{1}{(i)^{\sum r_k}} \frac{\partial^{\sum r_k}}{\partial u_1^{r_1} \ldots \partial u_m^{r_m}} \varphi_X(0),$$

for $\sum r_k \leq r$. As in the univariate case φ_X determines the d.f. F_X uniquely. Thus if B is the box $\{a_k < x \leq b_k, 1 \leq k \leq m\}$ then whenever $\mathbb{P}(X \in \partial B) = 0$ there holds

$$\mathbb{P}(X \in B) = \frac{1}{(2\pi)^m} \lim_{R\to\infty} \int_{-R}^{R} \cdots \int_{-R}^{R} \prod_{k=1}^{m} \frac{e^{-iu_k a_k} - e^{-iu_k b_k}}{iu_k} \varphi_X(u)\, du.$$

In particular if φ_X is integrable over $\mathbb{R}^m$ then F_X has a density f_X given by

$$f_X(x) = (2\pi)^{-m} \int \cdots \int_{\mathbb{R}^m} e^{-i\langle u, x\rangle} \varphi_X(u)\, du.$$

Exercises

1. (Parzen [45]) Let X, Y be bivariate normal with joint *pdf*

$$f_{XY}(x, y) = \frac{1}{2\pi\sigma_X\sigma_Y\sqrt{1-\rho^2}} \exp\left(-\frac{1}{2} Q(x, y)\right),$$

where

$$Q(x, y) = \frac{1}{1-\rho^2}\left[\left(\frac{x-\mu_X}{\sigma_X}\right)^2 - 2\rho\left(\frac{x-\mu_X}{\sigma_X}\right)\left(\frac{y-\mu_Y}{\sigma_Y}\right) + \left(\frac{y-\mu_Y}{\sigma_Y}\right)^2\right].$$

Compute

$$E(Y|X), \qquad \mathrm{Var}(Y|X), \qquad E(e^{iuY}|X).$$

2. (Breiman [5]) Prove Jensen's inequality: If $\varphi(y)$ is a convex function then

$$\varphi(E(Y|X)) \leq E(\varphi(Y)|X) \qquad \text{a.s.}$$

3. (Breiman [5]) Let $\Omega = [-1, +1]$, $\mathscr{F} = \mathscr{B}([-1, +1])$, $P(dx) = \frac{1}{2}\, dx$. If $X_1(x) = x^2$, $X_2(x) = x^4$ find

$$P(A|X_1), \qquad P(A|X_2),$$
$$E(Y|X_1), \qquad E(Y|X_2).$$

4. (Parzen [45]) Find $E(Y|X)$ when

(a) $$f_{XY}(x, y) = \begin{cases} 6xy(2-x-y), & 0 \leq x, y \leq 1 \\ 0 & \text{otherwise.} \end{cases}$$

(b) $$f_{XY}(x, y) = \begin{cases} 4y(x-y)e^{-(x+y)} & 0 \leq y \leq x < \infty \\ 0 & \text{otherwise.} \end{cases}$$

(c) $$f_{XY}(x, y) = \begin{cases} \frac{1}{8}(y^2 - x^2)e^{-y}, & 0 \leq |x| \leq y < \infty \\ 0 & \text{otherwise.} \end{cases}$$

(d) $$f_{XY}(x, y) = \frac{\sqrt{3}}{2\pi} e^{-(x^2+xy+y^2)}, \qquad -\infty < x, y < \infty$$

5. (Parzen [45]) A certain firm finds that the quantity X it sells of a certain item

is gamma distributed with parameters $(2, 2\mu)$, where μ itself is gamma distributed with parameters (α, β). Find the *pdf* of X.

6. (Parzen [45]) Let X, Y be independent Poisson. Find the conditional distribution of X given $X + Y$.

7. (Parzen [45]) Three players (denoted a, b, and c) take turns at playing a fair game according to the following rules. At the start a and b play, while c is out. The winner of the match between a and b plays c. The winner of the second match then plays the loser of the first match. The game continues in this way until a player wins twice in succession, thus becoming the winner of the game. Let A, B, and C denote, respectively, the events that a, b, or c is the winner of the game.

   (a) Find $P(A)$, $P(B)$, $P(C)$.
   (b) Find the mean duration of the game.

8. (Parzen [45]) The Thief of Baghdad has been placed in a dungeon with three doors. One of the doors leads into a tunnel that returns him to the dungeon after one day's travel through the tunnel. Another door leads to a similar tunnel (called the Long Tunnel) whose traversal requires three days rather than one day. The third door leads to freedom. Assume that the Thief is equally likely to choose each door (that is, each time he chooses a door he does not know what lies beyond). Find the mean number of days the Thief will be imprisoned from the moment he first chooses a door to the moment he chooses the door leading to freedom.

9. Let T be exponential with mean 1, $t > 0$. Compute

$$E(T \mid T \wedge t), \qquad E(T \mid T \vee t).$$

10. (Karlin and Taylor [31]) The random variables X and Y have the following properties:

    (a) X is positive a.s., and has a continuous *pdf* $f(x)$.
    (b) $Y|X$ is uniformly distributed on $(0, X)$.
    (c) Y and $X - Y$ are independent.

    Prove that X is Gamma $(2, \beta)$.

11. (Gnedenko [23]) There are 2^n tickets contained in a box; the number i $(i = 0, 1, \ldots, n)$ is marked on $\binom{n}{i}$ of them. m tickets are drawn at random and X is the sum of the numbers marked on them. Find $\mathbf{E}X$ and $\mathrm{Var}(X)$.

12. (Parzen [45]) Suppose

$$\varphi_X(u) = \exp\{\mu[e^{\lambda(e^{iu}-1)} - 1]\}$$

    Find $\mathbb{P}(X = 0)$, $\mathbf{E}X$, $\mathrm{Var}(X)$.

13. (Parzen [45]) A young man and a young lady plan to meet between 5 and 6 P.M., each agreeing not to wait more than 10 minutes for the other. Find the probability that they will meet if they arrive independently at random times between 5 and 6 P.M.

14. (Parzen [45]) In firing a missile at a given target, suppose that the vertical and horizontal components of the deviation of the impact from the center of the target can be regarded as normally and independently distributed about the center of the target with standard deviation of 3 miles. Suppose that one is aiming at a circular target with a diameter of 6 miles. What is the probability of the missile hitting the target? What is the smallest number of missiles that must be fired so that the probability is at least 0.99 of getting at least one hit?

15. (Parzen [45]) Let X and Y have joint probability density function

$$f_{X,Y}(x, y) = \frac{1}{\pi}, \qquad \text{if } x^2 + y^2 \leq 1.$$

Are X and Y (i) uncorrelated; (ii) independent?

16. (Parzen [45]) Show that the sum of n independent random variables, each obeying a gamma distribution with parameters r and λ, has the same distribution as the sum of rn independent exponentially distributed random variables with means $1/\lambda$.

17. (Parzen [45]) Find $\mathbb{P}(1 < X \leq 4)$ if

(i) $X = X_1 + X_2$
(ii) $X = \min(X_1, X_2)$

in which X_1 and X_2 are independent identically distributed random variables possessing the probability law described.

(a) X_1 and X_2 are $N(1, 2)$.
(b) X_1 and X_2 are Poisson, means 2.
(c) X_1 and X_2 are binomial, means 1 and variances 0.8.
(d) X_1 and X_2 are geometric, means 2.
(e) X_1 and X_2 are uniformly distributed over $[0, 3]$.

18. (Parzen [45]) Let X_1, X_2, X_3, X_4 be independent $N(0, 1)$. Let $a_1, a_2 > 0$. Evaluate $\mathbb{P}(R_1 > R_2)$ where

$$R_1 = a_1\sqrt{X_1^2 + X_2^2}, \qquad R_2 = a_2\sqrt{X_3^2 + X_4^2}.$$

Section III
Limit Laws

In Section II we dealt with finite families $X_1, \ldots, X_n$ of joint random variables. Now we shall be considering full sequences $X_1, X_2, \ldots$ of an infinity of joint random variables. The distribution of such a sequence is determined by the various finite-dimensional d.f.s $F_{X_{n_1}, \ldots, X_{n_k}}$; but the jump to infinity introduces many new considerations. In particular, we shall deal with limits, events that occur infinitely often (i.o.), tail events, and various modes of convergence.

Law of Large Numbers

The various forms of the Law of Large Numbers deal, in the most primitive case, with sums $S_n = X_1 + \cdots + X_n$ of independent identically distributed random variables $X_1, X_2, \ldots$ having finite expectations. These laws state that $\frac{S_n}{n}$ converges to $\mu = \mathbf{EX}_1$, in some mode of convergence. The Weak Law establishes *convergence in probability*; i.e. for any $\varepsilon > 0$

$$\lim_{n \to \infty} \mathbb{P}\left(\left|\frac{S_n}{n} - \mu\right| \geq \varepsilon\right) = 0.$$

The Strong Law establishes *a.s. convergence*; i.e.

$$\mathbb{P}\left(\lim_{n \to \infty} \frac{S_n}{n} = \mu\right) = 1.$$

Extensions of these laws involve relaxing the hypotheses that $X_1, X_2, \ldots$ be independent or identically distributed and can take a more general form $\frac{1}{b_n}(S_n - a_n) \to 0$, where $\{b_n\}$ is a positive sequence tending to ∞ and $\{a_n\}$ is appropriately defined. The skill in mastering these laws is to avoid making any additional assumptions on the X_is, such as the existence of second moments. For if we make, say, the assumption that the X_is have second moments, then the Weak Law is an immediate consequence of Chebyshev's Inequality, since $\mathrm{Var}\left(\frac{S_n}{n}\right) = \frac{\mathrm{Var}(X_1)}{n}$.

To prove the Strong Law we begin with two fundamental results.

Borel-Cantelli Lemma. *If* A_n *are events for which* $\sum_{n=1}^{\infty} \mathbb{P}(A_n) < \infty$ *then* $\mathbb{P}(A_n \text{ i.o}) = 0$. *Conversely if the events* A_n *are independent and* $\sum_{n=1}^{\infty} \mathbb{P}(A_n) = \infty$ *then* $\mathbb{P}(A_n \text{ i.o.}) = 1$.

PROOF. To prove the first half, observe that

$$\mathbb{P}(A_n \text{ i.o}) = \mathbb{P}\left(\lim_m \bigcup_{n=m}^{\infty} A_n\right) = \lim_m \mathbb{P}\left(\bigcup_{n=m}^{\infty} A_n\right) \le \lim_m \sum_{n=m}^{\infty} \mathbb{P}(A_n) = 0.$$

To prove the second half, observe that

$$\mathbb{P}\left(\bigcap_{n=m}^{\infty} A_n^c\right) = \prod_{n=m}^{\infty} [1 - \mathbb{P}(A_n)],$$

and since $\log(1-x) \le -x$ this product is zero for any m. □

Kolmogorov's Inequality. *Let* $X_1, \ldots, X_n$ *be independent random variables with finite second moments,* $\mathbf{E}X_k = 0$ *for all* k. *Set* $S_k = X_1 + \cdots + X_k$, $1 \le k \le n$. *For any* $\lambda > 0$

$$\mathbb{P}\left(\max_{1\le k\le n} |S_k| \ge \lambda\right) \le \frac{\mathbf{E}S_n^2}{\lambda^2}.$$

PROOF. Let T be the first k such that $|S_k| \ge \lambda$. (If no such k exists then set $T = n + 1$, say.) Then

$$\begin{aligned}
\mathbf{E}S_n^2 &\ge \sum_{k=1}^{n} \mathbf{E}I_{\{T=k\}} S_n^2 = \sum_{k=1}^{n} \mathbf{E}I_{\{T=k\}}[S_k^2 + 2S_k(S_n - S_k) + (S_n - S_k)^2] \\
&\ge \sum_{k=1}^{n} \mathbf{E}I_{\{T=k\}}[S_k^2 + 2S_k(S_n - S_k)] \\
&= \sum_{k=1}^{n} \mathbf{E}I_{\{T=k\}} S_k^2 \ge \lambda^2 \mathbb{P}(T \le n) = \lambda^2 \mathbb{P}\left(\max_{1\le k\le n} |S_k| \ge \lambda\right).
\end{aligned}$$
□

Observe that Kolmogorov's Inequality is a strengthening of Chebyshev's Inequality, which only asserts that

$$\mathbb{P}(|S_n| \geq \lambda) \leq \frac{\mathbf{E}S_n^2}{\lambda^2}.$$

Underlying Kolmogorov's Inequality is the phenomenon that for sums of independent random variables, if $\max_{1 \leq k \leq n} |S_k|$ is large then $|S_n|$ is probably large, too (e.g., the Reflection Principle).

Corollary I. *Let* $X_1, X_2, \ldots$ *be independent random variables with finite second moments,* $\mathbf{E}X_n = 0$ *for all* n. *If* $\sum_n \frac{\mathbf{E}X_n^2}{n^2} < \infty$ *then* $\frac{1}{n}\sum_{k=1}^{n} X_k \to 0$ a.s.

PROOF. Set $S_n = X_1 + \cdots + X_n$. It follows from Kolmogorov's Inequality that for any $\varepsilon > 0$ and any $k \geq 0$

$$\mathbb{P}\left(\frac{|S_n|}{n} \geq \varepsilon \text{ for some } 2^k \leq n < 2^{k+1}\right)$$

$$\leq \mathbb{P}\left(\max_{1 \leq n \leq 2^{k+1}} |S_n| \geq \varepsilon 2^k\right) \leq \frac{1}{\varepsilon^2 4^k} \sum_{n=1}^{2^{k+1}} \mathbf{E}X_n^2.$$

Since the right-hand side is summable over k it follows from the Borel-Cantelli lemma that $\mathbb{P}\left(\frac{|S_n|}{n} \geq \varepsilon \text{ i.o.}\right) = 0$. □

PROOF OF THE STRONG LAW. Define $\tilde{X}_n = X_n I_{\{|X_n| \leq n\}}$. By Property (P_4') of expectation and the first half of the Borel-Cantelli lemma it follows that $\mathbb{P}(\tilde{X}_n \neq X_n \text{ i.o.}) = \mathbb{P}(|X_n| > n \text{ i.o.}) = 0$. Furthermore

$$\lim_{n \to \infty} \mathbf{E}\tilde{X}_n = \lim_{n \to \infty} \mathbf{E}X_1 I_{\{|X_1| \leq n\}} = \mathbf{E}X_1.$$

Thus, by virture of the preceding corollary, it suffices to establish that $\sum_{n=1}^{\infty} \frac{\operatorname{Var}(\tilde{X}_n)}{n^2} < \infty$. Since $\operatorname{Var}(\tilde{X}_n) \leq \mathbf{E}\tilde{X}_n^2 = \mathbf{E}X_1^2 I_{\{|X_1| \leq n\}}$, we estimate

$$\sum_{n=1}^{\infty} \frac{\operatorname{Var}(\tilde{X}_n)}{n^2} \leq \sum_{n=1}^{\infty} \frac{1}{n^2} \int_{|x| \leq n} x^2 \, dF_X(x)$$

$$= \sum_{n=1}^{\infty} \sum_{k=1}^{n} \frac{1}{n^2} \int_{k-1 < |x| \leq k} x^2 \, dF_X(x) = \sum_{k=1}^{\infty} \sum_{n=k}^{\infty} \frac{1}{n^2} \int_{k-1 < |x| \leq k} x^2 \, dF_X(x)$$

$$\leq 2 \sum_{k=1}^{\infty} \frac{1}{k} \int_{k-1 < |x| \leq k} x^2 \, dF_X(x) \leq 2\mathbf{E}|X_1|.$$

Here F_X is the common d.f. of the X_ns, and in the next-to-last inequality we used

$$\sum_{n=k}^{\infty} \frac{1}{n^2} \leq \frac{2}{k}. \qquad \square$$

A basic result in probability is *Kolmogorov's Zero-One Law.* It states that *if $X_1, X_2, \ldots$ is an independent sequence of random variables then the probability of any tail event is either zero or one.* A *tail event* is any event concerning the X_ns that is determined by the tail $X_N, X_{N+1}, \ldots$ of the sequence alone, for any N. For example, tail events can be

$$\left\{\lim_{n\to\infty} \frac{1}{n} \sum_{k=1}^{n} X_k = \mu\right\}, \left\{\limsup_{n} \frac{X_{3n+2} - 5}{2^n} \leq 10\right\}$$

$$\left\{\sum_{n=1}^{\infty} X_n \text{ converges}\right\}, \left\{X_n \sim \frac{\binom{2n}{n}}{4^n} \text{ asymptotically}\right\}$$

Using this law, we can prove the converse to the Strong Law of Large Numbers, which states that if $\mathbf{E}|X_1| = \infty$ then $\frac{S_n}{n}$ diverges a.s. For if $\frac{S_n}{n}$ were to converge on a set of positive probability then, by the Zero-One Law, it would converge a.s. However, whenever $\frac{S_n}{n}$ converges, $\frac{X_n}{n}$ tends to zero. If this happens a.s. then $\mathbb{P}(|X_n| > n \text{ i.o.}) = 0$, and so by the second half of the Borel-Cantelli Lemma and Property (P_4') of expectation it would follow that $\mathbf{E}|X_1| < \infty$. Note also the following converse. If the X_ns are nonnegative and $\mathbf{E}X_1 = \infty$ then $\frac{S_n}{n} \to \infty$. Indeed

$$\frac{S_n}{n} \geq \frac{1}{n} \sum_{k=1}^{n} \min(X_k, T) \to \mathbf{E} \min(X_1, T),$$

and as $T \to \infty$ the right-hand side approaches $\mathbf{E}X_1$.

Weak Convergence

Let $\{F_n\}$ be a sequence of (univariate) d.f.s, corresponding to random variables $\{X_n\}$. We say that F_n *converges weakly*, or *in distribution*, to the d.f. F, denoted $F_n \xrightarrow{\mathscr{D}} F$, if $\lim_n F_n(x) = F(x)$ at all points x of continuity for F. The sequence $\{F_n\}$ is said to be *tight* if for any $\varepsilon > 0$ there is a bounded interval K such that $\mathbb{P}(X_n \notin K) \leq \varepsilon$, for all n. In order to study weakly convergent sequences, we need to enlarge the class of functions under consideration. Say that a real-valued function F defined on $\mathbb{R}$ is a *subdistribution function* (s.d.f.) if it is increasing and right-continuous, with $F(-\infty) \geq 0$, $F(\infty) \leq 1$. For s.d.f.s we still write $F_n \xrightarrow{\mathscr{D}} F$ whenever $\lim_n F_n(x) = F(x)$ at all points x of continuity for F.

Helly-Bray Theorem. *The set of sub-distribution functions is sequentially compact.*

PROOF. Let $\{F_n\}$ be a sequence of s.d.f.s, and let S be a countable dense set in $\mathbb{R}$. By the Cantor diagonalization process we can extract a subsequence $\{F_n'\}$ with the property that $G(y) = \lim_n F_n'(y)$ exists, for all $y \in S$. The limit function G, defined on S, is increasing. Now define F on all of $\mathbb{R}$ by

$$F(x) = \inf_{x < y \in S} G(y).$$

Then F is an s.d.f. To see that in fact $F_n' \xrightarrow{\mathcal{D}} F$ let $x \in \mathbb{R}$ and choose sequences $\{y_k\}$ and $\{z_k\}$ from S such that y_k strictly increases to x and z_k strictly decreases to x. For each k

$$F_n'(y_{k+1}) \leq F_n'(x) \leq F_n'(z_{k+1}), \qquad \forall n,$$

and thus

$$F(y_k) \leq G(y_{k+1}) \leq \liminf_n F_n'(x) \leq \limsup_n F_n'(x) \leq G(z_{k+1}) \leq F(z_k).$$

As $k \to \infty$, we arrive at the condition

$$F(x-) \leq \liminf_n F_n'(x) \leq \limsup_n F_n'(x) \leq F(x+). \qquad \square$$

Corollary II. *Let* $\{F_n\}$ *be a sequence of s.d.f.s. If there is a single s.d.f.* F *to which every weakly convergent subsequence* $\{F_n'\}$ *converges, then the full sequence* $F_n \xrightarrow{\mathcal{D}} F$.

PROOF. Let x be a point of continuity of F. By the Helly-Bray Theorem, every subsequence $\{F_n'\}$ contains a further subsequence $\{F_n''\}$ for which $F_n''(x) \to F(x)$. Thus $F_n(x) \to F(x)$. $\square$

Prohorov's Theorem. *If we restrict ourselves to d.f.s, then the closure of a set of d.f.s is sequentially compact (i.e., weak limits are themselves d.f.s) if and only if the set is tight.*

PROOF. Let $\{F_n\}$ be a tight sequence of d.f.s that converge weakly to an s.d.f. F. Given $\varepsilon > 0$ choose $a < b$ such that a and b are points of continuity for F and $F_n(b) - F_n(a) \geq 1 - \varepsilon$, for all n. Then $F(\infty) - F(-\infty) \geq F(b) - F(a) \geq 1 - \varepsilon$, and since ε is arbitrary we conclude that F is in fact a d.f. Conversely, suppose our set of d.f.s is not tight, and choose $\varepsilon > 0$ and a sequence $\{F_n\}$ such that $F_n(n) - F_n(-n) \leq 1 - \varepsilon$. Then if $\{F_n'\}$ is a weakly convergent subsequence, $F_n' \xrightarrow{\mathcal{D}} F$, it follows that $F(\infty) - F(-\infty) \leq 1 - \varepsilon$. $\square$

Theorem III. *Let* $\{F_n\}$ *be a sequence of d.f.s that converge weakly to a d.f.* F. *Then for any bounded continuous function* f *defined on* $\mathbb{R}$,

$$\int f(x)\, dF_n(x) \to \int f(x)\, dF(x).$$

PROOF. Let $\varepsilon > 0$ be given. Since $\{F_n\}$ is necessarily tight we can choose a bounded interval $K = [a, b]$ such that $F(b) - F(a) \geq 1 - \varepsilon$, and $F_n(b) - F_n(a) \geq 1 - \varepsilon$, for all n. Since f is *uniformly* continuous on K we can find a step function g defined on K whose partition points are all points of continuity for F, such that $\sup_{x \in K} |f(x) - g(x)| \leq \varepsilon$. It follows directly from the definition of weak convergence that for such a step function, whose partition points are points of continuity for F,

$$\lim_{n\to\infty} \int_K g\, dF_n = \int_K g\, dF.$$

Let M be a bound for $|f|$. Then

$$\begin{aligned}\left|\int f\, dF_n - \int f\, dF\right| &\leq \left|\int_{K^c} f\, dF_n - \int_{K^c} f\, dF\right| \\ &\quad + \left|\int_K (f - g)\, dF_n - \int_K (f - g)\, dF\right| + \left|\int_K g\, dF_n - \int_K g\, dF\right| \\ &\leq 2\varepsilon(M + 1) + \left|\int_K g\, dF_n - \int_K g\, dF\right|.\end{aligned}$$

Since ε was arbitrary, our result follows. □

Lemma IV. *There exists a constant* $0 < \alpha < \infty$ *such that for any random variable* X *with characteristic function* φ_X, *and for any* $u > 0$

$$\mathbb{P}\left(|X| > \frac{1}{u}\right) \leq \frac{\alpha}{u} \int_0^u [1 - \operatorname{Re} \varphi_X(v)]\, dv.$$

PROOF.

$$\begin{aligned}\frac{1}{u}\int_0^u [1 - \operatorname{Re} \varphi_X(v)]\, dv &= \frac{1}{u}\int_0^u (1 - \mathbf{E} \cos vX)\, dv \\ &= \mathbf{E}\left(1 - \frac{\sin uX}{uX}\right) \\ &\qquad \text{(by Result II from analysis, pg. 5)} \\ &\geq \min_{|t|\geq 1}\left(1 - \frac{\sin t}{t}\right)\mathbb{P}\left(|X| > \frac{1}{u}\right).\end{aligned}$$

□

The next result is one of the most far-reaching theorems in probability theory.

Continuity Theorem. *Let* $\{F_n\}$ *be a sequence of d.f.s with characteristic functions* $\{\varphi_n\}$. *Assume that* $\lim_n \varphi_n(u) = \varphi(u)$ *exists for all* u, *and that* φ *is continuous at* $u = 0$. *Then there is a d.f.* F *for which* $F_n \xrightarrow{\mathscr{D}} F$, *and* φ *is its characteristic function.*

PROOF. We first show that $\{F_n\}$ is tight. Let X_n be a random variable with d.f. F_n. Then by our previous lemma

$$\limsup_n \mathbb{P}\left(|X_n| > \frac{1}{u}\right) \le \frac{\alpha}{u} \limsup_n \int_0^u [1 - \operatorname{Re} \varphi_n(v)]\, dv$$
$$= \frac{\alpha}{u} \int_0^u [1 - \operatorname{Re} \varphi(v)]\, dv$$

(by Result *I* from analysis, pg. 5)

Since $\lim_{u \to 0} \varphi(u) = 1$,

$$\lim_{u \to 0} \limsup_n \mathbb{P}\left(|X_n| > \frac{1}{u}\right) = 0.$$

From this, it follows that $\{F_n\}$ is tight.

Take any weakly convergent subsequence $F_n' \xrightarrow{\mathscr{D}} F$. Since $\{F_n\}$ is tight, F must be a d.f. Furthermore, since e^{iux} is a bounded continuous function of x,

$$\varphi(u) = \lim_n \int e^{iux}\, dF_n(x) = \int e^{iux}\, dF(x),$$

and so φ is the characteristic function of F. Since φ uniquely determines F, every weakly convergent subsequence of $\{F_n\}$ must have this same limit F, and thus the full sequence $F_n \xrightarrow{\mathscr{D}} F$. □

Corollary V. *Let* $\{F_n\}$ *be a sequence of d.f.s with characteristic functions* $\{\varphi_n\}$, *and let* F *be a d.f. with characteristic function* φ. *If* $\lim_n \varphi_n(u) = \varphi(u)$ *for every* u *then* $F_n \xrightarrow{\mathscr{D}} F$.

Using the Continuity Theorem we get "instant" proofs of the Weak Law of Large Numbers, the Central Limit Theorem, and much more. In what follows, note that if the complex numbers c_n have the limit c, then

$$\lim_n \left(1 + \frac{c_n}{n}\right)^n = e^c. \tag{1}$$

PROOF OF THE WEAK LAW OF LARGE NUMBERS. Assume without loss of generality that $\mu = 0$. (Otherwise simply replace X_n with $X_n - \mu$.) Let

$Y_n = \frac{S_n}{n}$. Then

$$\varphi_{Y_n}(u) = \varphi^n\left(\frac{u}{n}\right),$$

where φ is the common characteristic function of the X_ns. Since X_n has finite expectation and $\mu = 0$, it follows from Property (P_4) of characteristic functions that

$$\varphi(u) = 1 + o(u).$$

Thus by (1)

$$\lim_n \varphi_{Y_n}(u) = 1,$$

and so by the Continuity Theorem, $Y_n \xrightarrow{\mathscr{D}} 0$. Since this weak limit turns out to be a constant, it follows that in fact we obtain convergence in probability. □

Central Limit Theorem. *Let* $X_1, X_2, \ldots$ *be independent identically distributed random variables having second moments and set* $S_n = X_1 + \cdots + X_n$. *Then*

$$\frac{S_n - n\mu}{\sigma\sqrt{n}} \xrightarrow{\mathscr{D}} N(0, 1),$$

where $\mu = \mathbf{E}X_n$ *and* $\sigma^2 = \operatorname{Var} X_n$.

PROOF. Assume without loss of generality that $\mu = 0$ and $\sigma = 1$. $\left(\text{Otherwise simply replace } X_n \text{ with } \frac{X_n - \mu}{\sigma}.\right)$ Let $Y_n = \frac{S_n}{\sqrt{n}}$. Then

$$\varphi_{Y_n}(u) = \varphi^n\left(\frac{u}{\sqrt{n}}\right),$$

where, as earlier, φ is the common characteristic function of the X_ns. Since X_n has a second moment and $\mu = 0$, $\sigma = 1$ it follows from Property (P_4) of characteristic functions that

$$\varphi(u) = 1 - \frac{1}{2}u^2 + o(u^2).$$

Thus by (1)

$$\lim_n \varphi_{Y_n}(u) = e^{-(1/2)u^2}.$$ □

Poisson Approximation to Binomial. *Let* X_n *be binomial with parameters* n *and* p_n. *If* $np_n \to \lambda$ *then* $X_n \xrightarrow{\mathscr{D}} X$, *where* X *is Poisson with parameter* λ.

PROOF. By (1)

$$\lim_n \varphi_{X_n}(u) = \lim_n [1 + p_n(e^{iu} - 1)]^n = \exp[\lambda(e^{iu} - 1)].$$

Now use the Continuity Theorem. □

Theorem VI. *Let* $X(\Delta t)$ *be negative binomial with parameters* r *and* $p(\Delta t) = \lambda\Delta t + o(\Delta t)$. *Then* $X(\Delta t)\Delta t \xrightarrow{\mathscr{D}} X$ *as* $\Delta t \to 0$, *where* X *is gamma-distributed, with parameters* r *and* λ.

PROOF.

$$\lim_{\Delta t \to 0} \frac{p(\Delta t)}{1 - [1 - p(\Delta t)]e^{iu\Delta t}} = \frac{\lambda}{\lambda - iu}.$$ □

We present now the Berry-Esseen estimate (from Feller [17]) for the rate of convergence in the Central Limit Theorem, in the case where the X_ns have finite third moments.

Berry-Esseen Theorem. *Let* $X_1, X_2, \ldots$ *be independent identically distributed random variables having finite third moments. Set* $\mu = \mathbf{E}X$, $\sigma^2 = \operatorname{Var} X$, $\gamma = \mathbf{E}\left|\frac{X - \mu}{\sigma}\right|^3$ *and* $S_n = X_1 + \cdots + X_n$. *Let*

$$\Phi(x) = \operatorname{erf}(x) = \frac{1}{\sqrt{2\pi}} \int_{-\infty}^{x} e^{-(1/2)z^2}\, dz$$

be the d.f. for the N(0, 1) *distribution. Then*

$$\sup_x \left| \mathbb{P}\left(\frac{S_n - n\mu}{\sigma\sqrt{n}} \le x \right) - \Phi(x) \right| \le 3 \frac{\gamma}{\sqrt{n}}.$$

The proof of this result is based upon the two lemmas that follow.

Smoothing Inequality. *Let* F *and* G *be d.f.s having finite expectations and respective characteristic functions* φ *and* ψ, *and suppose* G *is continuously differentiable with* $|dG/dx| \le M$. *Then for any* $T > 0$

$$\sup_x |F(x) - G(x)| \le \frac{2}{\pi} \int_0^T \frac{|\varphi(u) - \psi(u)|}{u}\, du + \frac{24M}{\pi T}. \tag{2}$$

PROOF. Observe first that $\int_{-\infty}^{\infty} |F(x) - G(x)|\, dx < \infty$ since F and G have finite expectations, by Property (P_4') of expectation. Integrating by parts,

$$\varphi(u) - \psi(u) = -iu \int_{-\infty}^{\infty} [F(x) - G(x)]e^{iux}\, dx,$$

and thus for any $a \in \mathbb{R}$, $T > 0$

$$\int_{-T}^{T} \frac{\varphi(u) - \psi(u)}{-iu} e^{-iua}(T - |u|)\, du$$
$$= \int_{-T}^{T} \int_{-\infty}^{\infty} [F(x + a) - G(x + a)] e^{iux}(T - |u|)\, dx\, du$$
$$= 2 \int_{-\infty}^{\infty} [F(x + a) - G(x + a)] \frac{1 - \cos Tx}{x^2} dx. \tag{3}$$

Next we estimate this last term in (3).

Set $\alpha = \sup_x |F(x) - G(x)|$. There must be a point $b \in \mathbb{R}$ such that either $F(b) - G(b) = \alpha$ or else $F(b-) - G(b) = -\alpha$. The analyses of these two cases are similar, so we treat only the latter. Set $\delta = \dfrac{\alpha}{2M}$ and choose now $a = b - \delta$. By the Mean-Value theorem,

$$F(x + a) - G(x + a) \leq F(b-) - [G(b) + (x - \delta)M]$$
$$= -M(x + \delta), \qquad \text{for all } |x| < \delta.$$

It follows that

$$\int_{-\delta}^{\delta} [F(x + a) - G(x + a)] \frac{1 - \cos Tx}{x^2} dx$$
$$\leq -M \int_{-\delta}^{\delta} (x + \delta) \frac{1 - \cos Tx}{x^2} dx = -\alpha \int_0^{\delta} \frac{1 - \cos Tx}{x^2} dx.$$

Thus we can estimate the full integral by

$$\int_{-\infty}^{\infty} [F(x + a) - G(x + a)] \frac{1 - \cos Tx}{x^2} dx$$
$$\leq \alpha \left(2 \int_{\delta}^{\infty} \frac{1 - \cos Tx}{x^2} dx - \int_0^{\delta} \frac{1 - \cos Tx}{x^2} dx \right)$$
$$= \alpha \left(3T \int_{\delta T}^{\infty} \frac{1 - \cos x}{x^2} dx - \frac{\pi T}{2} \right) \leq \alpha \left(\frac{6}{\delta} - \frac{\pi T}{2} \right),$$

using the fact that

$$\int_0^{\infty} \frac{1 - \cos x}{x^2} dx = \frac{\pi}{2}.$$

Combining this with (3) yields the desired estimate for α. □

Lemma VII. *Let* X *be a random variable with a finite third moment, and suppose* $\mathbf{E}\mathrm{X} = 0$, $\mathbf{E}\mathrm{X}^2 = 1$, *and* $\mathbf{E}|\mathrm{X}|^3 = \gamma$. *Then for all* $|\mathrm{u}| \leq \dfrac{4}{3\gamma}\sqrt{\mathrm{n}}$ *there* *holds*

$$\left|\varphi^n\left(\frac{u}{\sqrt{n}}\right) - e^{-(1/2)u^2}\right| \le \frac{\gamma|u|^3}{24\sqrt{n}}\left(\frac{3}{\sqrt{n}}|u| + 4\right)\exp\left(-\frac{5}{18}\frac{n-1}{n}u^2\right),$$

where φ *is the characteristic function of* X. *In particular for* $\mathrm{n} \ge 10$ $\left(\textit{and } |\mathrm{u}| \le \frac{4}{3\gamma}\sqrt{\mathrm{n}}\right)$

$$\left|\varphi^n\left(\frac{u}{\sqrt{n}}\right) - e^{-(1/2)u^2}\right| \le \frac{\gamma|u|^3}{24\sqrt{n}}(|u| + 4)e^{-(1/4)u^2}. \tag{4}$$

Proof. We use the fact that for any $x \in \mathbb{R}$, $m \in \mathbb{N}$

$$\left|e^{ix} - \sum_{k=0}^{m-1}\frac{(ix)^k}{k!}\right| \le \frac{|x|^m}{m!}. \tag{5}$$

This follows at once upon identifying the expression on the left inside the modulus, for $x > 0$, as

$$(i)^m \int_0^x \int_0^{x_1} \cdots \int_0^{x_{m-1}} e^{iu}\, dx_m \ldots dx_2\, dx_1.$$

Applying (5) to e^{iuX} with $m = 3$, we estimate

$$\left|\varphi(u) - 1 + \frac{1}{2}u^2\right| \le \frac{1}{6}\gamma|u|^3; \tag{6}$$

thus if $|u| \le \frac{4}{3\gamma} < \sqrt{2}$ (since $\gamma \ge 1$) we have

$$|\varphi(u)| \le 1 - \frac{1}{2}u^2 + \frac{1}{6}\gamma|u|^3 \le 1 - \frac{5}{18}u^2 \le e^{-(5/18)u^2}. \tag{7}$$

Similarly, using the inequality

$$e^{-x} \le 1 - x + \frac{1}{2}x^2, \qquad \text{for all } x > 0,$$

we estimate from (6) again

$$\begin{aligned} |\varphi(u) - e^{-(1/2)u^2}| &\le \left|\varphi(u) - 1 + \frac{1}{2}u^2\right| + \left|e^{-(1/2)u^2} - 1 + \frac{1}{2}u^2\right| \\ &\le \frac{1}{6}\gamma|u|^3 + \frac{1}{8}u^4. \end{aligned} \tag{8}$$

Finally we note that the familiar factorization for $x^n - y^n$ leads to the inequality

$$|x^n - y^n| \le n|x - y|a^{n-1}$$

where $a = \max(|x|, |y|)$. Applying this with $x = \varphi\left(\frac{u}{\sqrt{n}}\right)$ and $y = \exp\left(-\frac{1}{2}\frac{u^2}{n}\right)$ and using our estimates (7) and (8) with u replaced by $\frac{u}{\sqrt{n}}$ leads straightway to our desired estimate. □

PROOF OF THE BERRY-ESSEEN THEOREM. Assume without loss of generality that $\mu = 0$ and $\sigma = 1$. $\left(\text{Otherwise simply replace } X_n \text{ with } \frac{X_n - \mu}{\sigma}.\right)$ In the Smoothing Inequality take F to be the d.f. of $\frac{S_n}{\sqrt{n}}$ and take G to be the d.f. of the $N(0, 1)$ distribution. This d.f. G is certainly continuously differentiable, and M can be taken as $\frac{1}{\sqrt{2\pi}}$. The respective characteristic functions are given by

$$\varphi(u) = \varphi_X^n\left(\frac{u}{\sqrt{n}}\right), \qquad \Psi(u) = e^{-(1/2)u^2}$$

where φ_X is the common characteristic function of the X_ns. Since $\gamma \geq 1$ the Berry-Esseen estimate is trivially true for $\sqrt{n} \leq 3$, so we may assume $n \geq 10$. Choose $T = \frac{4}{3\gamma}\sqrt{n}$ now so that we may conveniently apply our estimate (4). The combined estimate works out to be

$$\sup_x |F(x) - G(x)| \leq \frac{\gamma}{\pi\sqrt{n}}\left[\frac{1}{12}\int_0^T (u^3 + 4u^2)e^{-(1/4)u^2}\,du + 9\sqrt{\frac{2}{\pi}}\right].$$

Now observe that

$$\int_0^\infty (u^3 + 4u^2)e^{-(1/4)u^2}\,du = 8(1 + \sqrt{\pi}).$$ □

To see that the $\frac{1}{\sqrt{n}}$ rate of the Berry-Esseen Theorem cannot be improved upon, consider the case of coin tossing, where S_n is binomial with parameters n and $p = \frac{1}{2}$. Then whenever n is even

$$\mathbb{P}\left(S_n - \frac{n}{2} \leq 0\right) - \frac{1}{2} = \frac{\binom{n}{n/2}}{2^{n+1}}.$$

and by Stirling's approximation the right-hand side is asymptotically $\frac{1}{\sqrt{2\pi n}}$.

The notions of weak convergence and tightness, the Continuity Theorem, the Laws of Large Numbers, the Central Limit Theorem, and many other limit laws carry over to the multivariate setting as well. We say that a sequence $X_1, X_2, \ldots$ of random vectors in $\mathbb{R}^m$ converges weakly, or in distribution to X, denoted $X_n \xrightarrow{\mathscr{D}} X$, if for any Borel set $B \subseteq \mathbb{R}^m$ with $\mathbb{P}(X \in \partial B) = 0$,

$$\lim_n \mathbb{P}(X_n \in B) = \mathbb{P}(X \in B).$$

We say that the sequence is tight if for any $\varepsilon > 0$ there exists a compact set $K \subseteq \mathbb{R}^m$ such that

$$\mathbb{P}(X_n \notin K) \leq \varepsilon, \qquad \text{for all } n.$$

Many of our results in the univariate case hold as well in the multivariate setting. In particular if $X_n \xrightarrow{\mathscr{D}} X$ then $\mathbf{E}f(X_n) \to \mathbf{E}f(X)$ for every bounded continuous function f defined on $\mathbb{R}^m$. Prohorov's Theorem is valid, and if $\{X_n\}$ is tight then it has a weakly convergent subsequence (which converges to some random vector X). The joint characteristic function of a distribution uniquely determines that distribution, and the Continuity Theorem continues to hold. Thus *if* $\{X_n\}$ *is a sequence of random vectors in* $\mathbb{R}^m$ *and if* X *is a random vector in* $\mathbb{R}^m$, *then* $X_n \xrightarrow{\mathscr{D}} X$ *if and only if* $\varphi_{X_n}(u) \to \varphi_X(u)$ *for every* $u \in \mathbb{R}^m$. The proofs of these results are straightforward adaptations of those from the univariate case. In particular, the Helly-Bray Theorem is proved by getting a subsequence that converges at all points in $\mathbb{R}^m$ with rational coordinates. The proof of the Continuity Theorem is in fact reduced to the univariate case by simply observing that a sequence $\{X_n\}$ of random vectors is tight if and only if the sequences $\{X_{n,i}\}$ of the ith marginals are tight. We present now the multivariate Central Limit Theorem.

Multivariate Central Limit Theorem. *Let* $X_1, X_2, \ldots$ *be an independent identically distributed sequence of random vectors in* $\mathbb{R}^m$ *with finite second moments,* $\mathbf{E}\|X_n\|^2 < \infty$. *Let* μ *be the common mean of the* X_ns and let Γ *be the common* $m \times m$ *covariance matrix. Set* $S_n = X_1 + \cdots + X_n$. *Then*

$$\frac{S_n - n\mu}{\sqrt{n}} \xrightarrow{\mathscr{D}} N(0, \Gamma).$$

Proof. Assume without loss of generality that $\mu = 0$. (Otherwise simply replace X_n with $X_n - \mu$.) Set $Y_n = \dfrac{S_n}{\sqrt{n}}$. Then

$$\varphi_{Y_n}(u) = \varphi^n\left(\frac{u}{\sqrt{n}}\right),$$

where φ is the common characteristic function of the X_ns. Given $u \in \mathbb{R}^m$, $u \neq 0$, let $\hat{u}$ denote the unit vector $\dfrac{u}{\|u\|}$. Since $\langle \hat{u}, X \rangle$ has a finite second moment, it follows from Property (P_4) of characteristic functions that

$$\varphi(u) = \varphi_{\langle \hat{u}, X_1 \rangle}(\|u\|) = 1 - \frac{1}{2}\mathbf{E}(\langle u, X_1 \rangle)^2 + o(\|u\|^2).$$

Thus by (1)

$$\lim_n \varphi_{Y_n}(u) = \exp\left[-\frac{1}{2}\mathbf{E}(\langle u, X_1 \rangle)^2\right] = \exp\left(-\frac{1}{2}\langle \Gamma u, u \rangle\right).$$

Now apply the Continuity Theorem. □

Bochner's Theorem (from Loève [40])

Let $S \subseteq \mathbb{R}$ be closed under subtraction. Say that a complex-valued function ψ defined on S is nonnegative definite if for any $n \in \mathbb{N}$, any $u_1, \ldots, u_n \in S$ and any $\xi_1, \ldots, \xi_n \in \mathbb{C}$ there holds

$$\sum_{j=1}^{n} \sum_{k=1}^{n} \psi(u_j - u_k)\xi_j \bar{\xi}_k \geq 0. \tag{9}$$

We have already seen (Property (P_7) of characteristic functions) that every characteristic function is nonnegative definite. Bochner's theorem is the converse to this, and as such characterizes characteristic functions completely.

Herglotz Lemma. *Let* $\mathrm{S} = \{\ldots, -2\mathrm{d}, -\mathrm{d}, 0, \mathrm{d}, 2\mathrm{d}, \ldots\}$ *where* $\mathrm{d} > 0$, *and suppose* $\psi(0) = 1$. *Then* ψ *defined on* S *is nonnegative definite if and only if it coincides on* S *with the characteristic function of a random variable supported in* $\left[-\frac{\pi}{\mathrm{d}}, \frac{\pi}{\mathrm{d}}\right]$.

Proof. Assume without loss of generality that $d = 1$. $\left(\text{Otherwise replace } \psi(u) \text{ with } \psi\left(\frac{u}{d}\right).\right)$ Set

$$f_n(x) = \frac{1}{2\pi} \sum_{k=-n+1}^{n} \left(1 - \frac{|k|}{n}\right)\psi(k)e^{ikx},$$

and observe that

$$f_n(x) = \frac{1}{2\pi n} \sum_{j=1}^{n} \sum_{m=1}^{n} \psi(j - m)e^{-ijx}e^{imx}.$$

Thus if ψ is nonnegative definite, $f_n \geq 0$. Furthermore

$$\left(1 - \frac{|k|}{n}\right)\psi(k) = \int_{-\pi}^{\pi} f_n(x)e^{ikx}\,dx.$$

In particular since $\psi(0) = 0$, f_n integrates to one, and can thus be taken as the density for an absolutely continuous d.f. F_n supported in $[-\pi, \pi]$. Since

$$\psi(k) = \lim_n \int_{-\pi}^{\pi} e^{ikx}\, dF_n(x),$$

our result follows from an application of the Helly-Bray Theorem. □

Bochner's Theorem. *A complex-valued function ψ defined on $\mathbb{R}$ with $\psi(0) = 1$ is the characteristic function of some random variable if and only if it is continuous at the origin and nonnegative definite.*

PROOF. Assume that ψ is continuous at the origin and nonnegative definite. We first point out that ψ is necessarily uniformly continuous on $\mathbb{R}$. By considering (9) when $n = 2$ and $u_2 = 0$ we see that necessarily $\psi(-u) = \overline{\psi(u)}$ and $|\psi(u)| \le 1$, for all $u \in \mathbb{R}$. By considering (9) when $n = 3$ and $u_3 = 0$ we see that the Hermitian matrix

$$\begin{bmatrix} 1 & \psi(u) & \psi(v) \\ \overline{\psi(u)} & 1 & \psi(u-v) \\ \overline{\psi(v)} & \overline{\psi(u-v)} & 1 \end{bmatrix}$$

is nonnegative definite. Since its determinant is nonnegative we find that

$$\begin{aligned} |\psi(u) - \psi(v)|^2 &\le 1 - |\psi(u-v)|^2 - 2\,\mathrm{Re}\{\psi(u)\overline{\psi(v)})[1 - \psi(u-v)]\} \\ &\le 4|1 - \psi(u-v)|. \end{aligned}$$

Thus as long as ψ is continuous at the origin, it will be uniformly continuous on $\mathbb{R}$.

For every $T > 0$ and $x \in \mathbb{R}$

$$f_T(x) = \frac{1}{T}\int_0^T \int_0^T \psi(u-v)e^{-i(u-v)x}\, du\, dv \ge 0.$$

This is because the integral can be obtained as a limit of Riemann sums, since ψ is continuous; each Riemann sum is nonnegative, since ψ is nonnegative definite. The preceding integral transforms to

$$f_T(x) = \int_{-T}^{T} \left(1 - \frac{|u|}{T}\right)\psi(u)e^{-iux}\, du,$$

and thus for any $N > 0$

$$\begin{aligned} &\frac{1}{2\pi}\int_{-N}^{N} \left(1 - \frac{|x|}{N}\right) f_T(x)e^{iux}\, dx \\ &= \frac{N}{\pi}\int_{-T}^{T} \frac{1 - \cos N(u-v)}{[N(u-v)]^2}\left(1 - \frac{|v|}{T}\right)\psi(v)\, dv \\ &= \frac{1}{\pi}\int_{N(u-T)}^{N(u+T)} \frac{1 - \cos z}{z^2}\left(1 - \frac{|u - z/N|}{T}\right)\psi\left(u - \frac{z}{N}\right) dz. \end{aligned}$$

For any $N > 0$, $T > 0$ the left hand side of this equality is, up to normalization, a characteristic function. The right-hand side of this equality tends, as first $N \to \infty$ then $T \to \infty$, to $\psi(u)$. Thus, by the Continuity Theorem, ψ itself must be a characteristic function. □

Extremes (from Leadbetter, Lindgren and Rootzén [39])

In what follows let $\{X_n\}$ be an independent identically distributed sequence of random variables with common d.f. F. Set $M_n = \max(X_1, \ldots, X_n)$, $m_n = \min(X_1, \ldots, X_n)$ and let $M_n^{(k)}$ denote the kth largest from among $X_1, \ldots, X_n$. We shall be concerned with joint asymptotic distributions of these extremes.

Theorem VIII. *Let* $0 \le \tau \le \infty$ *and suppose that* $\{x_n\}$ *is a sequence of real numbers such that*

$$\lim_n n[1 - F(x_n)] = \tau. \tag{10}$$

Then

$$\lim_n \mathbb{P}(M_n \le x_n) = e^{-\tau}. \tag{11}$$

Conversely if (11) *holds for some* τ, $0 \le \tau \le \infty$, *then so does* (10).

PROOF. Observe that

$$\mathbb{P}(M_n \le x_n) = F^n(x_n) = \{1 - [1 - F(x_n)]\}^n. \tag{12}$$

Suppose first that $\tau < \infty$. If (10) holds then (11) follows at once from (1). Conversely if (11) holds then we must have $1 - F(x_n) \to 0$. Otherwise some subsequence $1 - F(x_{n_l})$ would be bounded away from zero, and by (12) then $\mathbb{P}(M_{n_l} \le x_{n_l}) \to 0$. By taking logarithms now in (11) and (12) we find that

$$-n[1 - F(x_n)][1 + o(1)] = n \log\{1 - [1 - F(x_n)]\} \to -\tau,$$

thus establishing (10). Finally if $\tau = \infty$ and (11) does not hold then we can extract a subsequence such that $\mathbb{P}(M_{n_l} \le x_{n_l}) \to e^{-\tau'}$ for some $\tau' < \infty$. But then by our argument for the case of τ finite, this would lead to the conclusion $n_l[1 - F(x_{n_l})] \to \tau'$, and so (10) does not hold. Similarly if $\tau = \infty$ then (11) also implies (10). □

Denote by x_F the right endpoint of the support of the X_ns,

$$x_F = \sup(x\colon F(x) < 1).$$

Corollary IX. (i) $M_n \to x_F$ a.s.
(ii) *If* $x_F < \infty$ *and* $F(x_F-) < 1$ *then any limit* $\lim_n \mathbb{P}(M_n \le x_n)$ *must equal either zero or one.*

PROOF. (i) If $x < x_F$ then (10) holds with $x_n \equiv x$ and $\tau = \infty$, and thus $\mathbb{P}(M_n \leq x) \to 0$. Since $\mathbb{P}(M_n > x_F) = 0$ for all n it follows that $M_n \to x_F$ in probability. Since $\{M_n\}$ is monotone it converges a.s., and hence $M_n \to x_F$ a.s. (ii) Suppose $\mathbb{P}(M_n \leq x_n) \to e^{-\tau}$. If $x_n < x_F$ i.o. then since $1 - F(x_n) \geq 1 - F(x_{F}-) > 0$ for these n, we have $\tau = \infty$. Otherwise, if $x_n \geq x_F$ for all n sufficiently large then $1 - F(x_n) = 0$ for these n, and we have $\tau = 0$. □

Regarding the minimum m_n, we have a parallel result that can be proved in an analogous fashion, since

$$\mathbb{P}(m_n > y_n) = [1 - F(y_n)]^n.$$

Theorem X. *Let* $0 \leq \eta \leq \infty$ *and suppose that* $\{y_n\}$ *is a sequence of real numbers such that*

$$\lim_n nF(y_n) = \eta. \tag{13}$$

Then

$$\lim_n \mathbb{P}(m_n > y_n) = e^{-\eta}. \tag{14}$$

Conversely if (14) *holds for some* η, $0 \leq \eta \leq \infty$, *then so does* (13).

Theorem XI. *Suppose the sequences* $\{x_n\}$ *and* $\{y_n\}$ *satisfy* (10) *and* (13), *respectively. Then the events* $\{M_n \leq x_n\}$ *and* $\{m_n > y_n\}$ *are asymptotically independent, so that*

$$\lim_n \mathbb{P}(M_n \leq x_n, m_n > y_n) = e^{-(\tau+\eta)}.$$

PROOF. Observe that

$$\begin{aligned}\mathbb{P}(M_n \leq x_n, m_n > y_n) &= \mathbb{P}(y_n < X_i \leq x_n; 1 \leq i \leq n)\\ &= [F(x_n) - F(y_n)]^n = \{1 - F(y_n) - [1 - F(x_n)]\}^n.\end{aligned}$$

If τ, $\eta < \infty$ then use (1). The cases where τ or η is infinite are dealt with simply since if, for example, $\tau = \infty$, then $\mathbb{P}(M_n \leq x_n, m_n > y_n) \leq \mathbb{P}(M_n \leq x_n) \to 0$. □

Regarding the rate of convergence in (11) we present the following bound.

Theorem XII. *Put* $\tau_n = n[1 - F(x_n)]$. *Then*

$$0 \leq e^{-\tau_n} - \left(1 - \frac{\tau_n}{n}\right)^n \leq \frac{\tau_n^2 e^{-\tau_n}}{2(n-1)} \leq \frac{0.3}{n-1},$$

and the first bound is asymptotically sharp, in the sense that if $\tau_n \to \tau$ *then*

$e^{-\tau_n} - \left(1 - \frac{\tau_n}{n}\right)^n \sim \frac{\tau^2 e^{-\tau}}{2n}$. *Furthermore if* $\tau - \tau_n \leq \log 2$ *then*

$$e^{-\tau_n} - e^{-\tau} = e^{-\tau}[\tau - \tau_n + \theta(\tau - \tau_n)^2],$$

where $0 < \theta < 1$.

This result follows from the lemma we provide.

Lemma XIII. (i) *If* $0 \leq x \leq n$ *then for all* n

$$0 \leq e^{-x} - \left(1 - \frac{x}{n}\right)^n \leq \frac{x^2 e^{-x}}{2(n-1)} \leq \frac{2e^{-2}}{n-1} \leq \frac{0.3}{n-1}$$

and

$$e^{-x} - \left(1 - \frac{x}{n}\right)^n = \frac{x^2 e^{-x}}{2n}\left[1 + O\left(\frac{1}{n}\right)\right] \text{ as } n \to \infty.$$

(ii) *If* $x - y \leq \log 2$ *then*

$$e^{-y} - e^{-x} = e^{-x}[x - y + \theta(x - y)^2],$$

with $0 < \theta < 1$.

Proof. (i) The first inequality is a consequence of the estimate $e^{-(x/n)} \leq 1 - \frac{x}{n}$. Next set

$$f(x) = \frac{x^2}{2(n-1)} - 1 + e^x\left(1 - \frac{x}{n}\right)^n.$$

Observe that

$$f'(x) = \frac{x}{n-1}\left[1 - e^x\left(1 - \frac{x}{n}\right)^{n-1}\left(1 - \frac{1}{n}\right)\right],$$

and the expression in brackets assumes its minimum in $[0, n]$ for $x = 1$. Thus $f' \geq 0$, and since $f(0) = 0$ our first result follows. The second result here follows from Taylor's expansion applied to $f_n(x) = 1 - e^x\left(1 - \frac{x}{n}\right)^n$, since $f_n(0) = f_n'(0) = 0$, $f_n''(0) = \frac{1}{n}$, and $f_n'''(x) = O\left(\frac{1}{n^2}\right)$ uniformly for x in bounded intervals.

(ii) By Taylor's theorem

$$e^{x-y} - 1 = x - y + \frac{1}{2}(x - y)^2 e^{\theta(x-y)},$$

where $0 < \theta < 1$. □

Given a sequence of real numbers $\{x_n\}$ we let S_n denote the number of exceedances of x_n by $X_1, \ldots, X_n$; i.e. the number of times $X_i > x_n$, $1 \le i \le n$. Clearly S_n is binomial with parameters n and $p_n = 1 - F(x_n)$. Thus by our Poisson Approximation to Binomial Theorem, if (10) holds with $0 \le \tau < \infty$ then S_n converges weakly to the Poisson distribution with parameter τ. (When $\tau = 0$ this is just the constant 0.) Furthermore if (10) holds with $\tau = \infty$ then $S_n \to \infty$ in probability, in the sense that $\lim_n \mathbb{P}(S_n \le k) = 0$, for all k. To argue this last point simply observe that for any $\theta \le np_n$,

$$\mathbb{P}(S_n \le k) \le \sum_{j=0}^{k} \binom{n}{j}\left(\frac{\theta}{n}\right)^j\left(1 - \frac{\theta}{n}\right)^{n-j} \to e^{-\theta} \sum_{j=0}^{k} \frac{\theta^j}{j!},$$

since the sum in the middle is a decreasing function of θ.

The extreme $M_n^{(k)}$ is intimately tied to S_n, since the events $\{M_n^{(k)} \le x_n\}$ and $\{S_n < k\}$ are identical. We are thus led to the following conclusion.

Theorem XIV. *Let* $0 \le \tau \le \infty$ *and suppose that* $\{x_n\}$ *is a sequence of real numbers such that* (10) *holds. Then for each* k

$$\lim_n \mathbb{P}(M_n^{(k)} \le x_n) = e^{-\tau} \sum_{j=0}^{k-1} \frac{\tau^j}{j!}. \tag{15}$$

Conversely if (15) *holds for some fixed* k, *then* (10) *holds, and hence* (15) *holds for all* k.

PROOF. We need only prove the converse. If (10) does not hold, then we can choose a subsequence such that $n_l[1 - F(u_{n_l})] \to \tau'$ where $\tau' \ne \tau$. Our previous argument then shows that $\mathbb{P}(S_n < k) \to e^{-\tau'} \sum_{j=0}^{k-1} \frac{(\tau')^j}{j!}$, which is different than the right-hand side of (15), since the function $e^{-x} \sum_{j=0}^{k-1} \frac{x^j}{j!}$ is strictly decreasing in $0 \le x \le \infty$. □

Theorem XV. *Let the levels* $x_n^{(1)} \ge \cdots \ge x_n^{(r)}$ *satisfy*

$$\lim_n n[1 - F(x_n^{(l)})] = \tau_l, \qquad 1 \le l \le r,$$

where $0 \le \tau \le \cdots \le \tau_r \le \infty$, *and define* $S_n^{(l)}$ *to be the number of exceedances of* $x_n^{(l)}$ *by* $X_1, \ldots, X_n$. *Then* $S_n^{(l)} - S_n^{(l-1)}$, $1 \le l \le r$, *are asymptotically independent, so that for any* $k_1, \ldots, k_r \ge 0$

$$\lim_n \mathbb{P}(S_n^{(l)} - S_n^{(l-1)} = k_l, 1 \le l \le r) = e^{-\tau_r} \prod_{l=1}^{r} \frac{(\tau_l - \tau_{l-1})^{k_l}}{k_l!}.$$

(Here we have adopted the convention that $S_n^{(0)} = \tau_0 = 0$.)

PROOF. The joint distribution of $(S_n^{(l)} - S_n^{(l-1)})_{l=1}^r$ is multinomial with parameters n and $p_{nl} = F(x_n^{(l-1)}) - F(x_n^{(l)})$, $1 \le l \le r$, where we have adopted the convention that $F(x_n^{(0)}) = 1$. Their joint characteristic function is thus

$$\varphi_n(u) = [1 + \sum_{l=1}^{r} p_{nl}(e^{iu_l} - 1)]^n.$$

By (1)

$$\lim_n \varphi_n(u) = \exp[\sum_{l=1}^{r} (\tau_l - \tau_{l-1})(e^{iu_l} - 1)],$$

which indeed corresponds to the product of the characteristic functions of Poissons with parameters $\tau_l - \tau_{l-1}$. □

Writing

$$\{M_n^{(1)} \le x_n^{(1)}, \ldots, M_n^{(r)} \le x_n^{(r)}\} = \{S_n^{(1)} = 0, S_n^{(2)} \le 1, \ldots, S_n^{(r)} \le r - 1\}$$

we see how to obtain the joint asymptotic distribution of the r largest maxima from that of the $S_n^{(l)}$s. Regarding rates of convergence Leadbetter, Lindgren, and Rootzén [39] establish that for $\tau_n = n[1 - F(x_n)]$,

$$\left| \mathbb{P}(M_n^{(k)} \le x_n) - e^{-\tau_n} \sum_{j=0}^{k-1} \frac{\tau_n^j}{j!} \right| \le \frac{\tau_n^2}{n},$$

and

$$\left| \mathbb{P}(M_n^{(k)} \le x_n) - e^{-\tau} \sum_{j=0}^{k-1} \frac{\tau^j}{j!} \right| \le \frac{\tau_n^2}{n} + |\tau_n - \tau|.$$

Finally, in connection with (10) we note that if the X_ns have a finite rth moment then

$$\lim_n n^r[1 - F(n)] = 0,$$

and thus by (11)

$$\frac{1}{\sqrt[r]{n}} \max(|X_1|, \ldots, |X_n|) \xrightarrow{\mathscr{D}} 0.$$

Examples

Let the X_ns be exponentially distributed, with d.f. $F(x) = 1 - e^{-x}$. Then

$$\lim_n \mathbb{P}(M_n \le x + \log n) = \exp(-e^{-x}).$$

If the X_ns have the Pareto distribution, with d.f. $F(x) = 1 - Kx^{-\alpha}$,

$x \geq K^{1/\alpha}$, then for $x > 0$

$$\lim_n \mathbb{P}(M_n \leq (Kn)^{1/\alpha}x) = \exp(-x^{-\alpha}).$$

For the uniform distribution, $F_n(x) = x$, $0 \leq x \leq 1$, we have for $x \leq 0$

$$\lim_n \mathbb{P}\left(M_n \leq 1 + \frac{x}{n}\right) = e^x.$$

For the Cauchy distribution, $F(x) = \dfrac{1}{2} + \dfrac{1}{\pi}\tan^{-1}x$, we have for $x > 0$

$$\lim_n \mathbb{P}\left(M_n \leq \frac{nx}{\pi}\right) = e^{-(1/x)}.$$

For the normal distribution, $F(x) = \operatorname{erf}(x)$, we have

$$\lim_n \mathbb{P}\left(M_n \leq \frac{1}{\sqrt{2\log n}}\left(x + \log\frac{n^2}{2\sqrt{\pi\log n}}\right)\right) = \exp(-e^{-x}).$$

(Use the fact that

$$1 - \operatorname{erf}(x) \sim \frac{1}{\sqrt{2\pi}}\frac{e^{-(1/2)x^2}}{x} \text{ as } x \to \infty.\Big)$$

Extremal Distributions

The d.f.s F and G are of the *same type* if there exist constants $a > 0$ and b such that $F(ax + b) = G(x)$, $\forall x$. A d.f. is *degenerate* if it has the form δ_y for some $y \in \mathbb{R}$; otherwise it is *nondegenerate*.

Khinchine's Convergence of Types Theorem. *Suppose that* $F_n \xrightarrow{\mathscr{D}} F$ *and* $F_n(a_n x + b_n) \xrightarrow{\mathscr{D}} G(x)$ *where* $a_n > 0$ *and* F *and* G *are nondegenerate. Then there exist* $a > 0$ *and* b *such that* $a_n \to a$, $b_n \to b$ *and* $G(x) = F(ax + b)$.

We prove Khinchine's Theorem with the help of the following lemma.

Lemma XVI. *Let* $a_n, a > 0$ *and suppose* $F_n \xrightarrow{\mathscr{D}} F$.

i) *If* $a_n \to a$ *and* $b_n \to b$ *then* $F_n(a_n x + b_n) \xrightarrow{\mathscr{D}} F(ax + b)$.
(ii) *If* $a_n \to \infty$ *then* $F_n(a_n x) \xrightarrow{\mathscr{D}} \delta_x$.
(iii) *If* $\{b_n\}$ *is unbounded then* $F_n(x + b_n)$ *cannot converge weakly.*
(iv) *If* $F_n(a_n x + b_n) \xrightarrow{\mathscr{D}} G(x)$ *and if* F *and* G *are nondegenerate then*

$$0 < \inf_n a_n \leq \sup_n a_n < \infty, \qquad \sup_n |b_n| < \infty.$$

(v) *If* $\mathrm{F(x) = F(ax + b)}$ *for all* x, *and if* F *is nondegenerate, then* $\mathrm{a} = 1$ *and* $\mathrm{b} = 0$.

PROOF. (i) Let x be a point of continuity for $F(ax + b)$ and let $\varepsilon > 0$. Choose continuity points y, z for F so that $y < ax + b < z$ and $F(z) - F(y) < \varepsilon$. For n sufficiently large $y < a_n x + b_n < z$, $|F_n(y) - F(y)| < \varepsilon$ and $|F_n(z) - F(z)| < \varepsilon$. Thus

$$F(ax + b) - 2\varepsilon < F(y) - \varepsilon < F_n(y) \leq F_n(a_n x + b_n)$$
$$\leq F_n(z) < F(z) + \varepsilon < F(ax + b) + 2\varepsilon.$$

(ii) Given $\varepsilon > 0$ choose a point of continuity y for F large enough so that $F(y) > 1 - \varepsilon$. If $x > 0$ then for all n sufficiently large $a_n x > y$ and $|F_n(y) - F(y)| < \varepsilon$, so that $F_n(a_n x) \geq F_n(y) > F(y) - \varepsilon > 1 - 2\varepsilon$. Thus $\lim_n F_n(a_n x) = 1$ for $x > 0$. Similarly $\lim_n F_n(a_n x) = 0$ for $x < 0$.

(iii) Suppose $b_n' \to \infty$ for some subsequence. (The case where some subsequence tends to $-\infty$ is handled similarly.) Given $\varepsilon > 0$ choose a point of continuity y for F so that $F(y) > 1 - \varepsilon$. For any x eventually $x + b_n' > y$ and $F_n'(y) > 1 - 2\varepsilon$, so that $F_n'(x + b_n') > 1 - 2\varepsilon$. Thus if $F_n(x + b_n)$ were to converge weakly to G, then necessarily $G(x) = 1$ for all continuity points of G, which is impossible.

(iv) If $\{a_n\}$ is not bounded above then take a subsequence $a_n' \to \infty$. By (ii) it follows that $F_n'(a_n' x) \xrightarrow{\mathscr{D}} \delta_x$, and by applying (iii) to the d.f.s $F_n'(a_n' x)$ it follows further that $\{b_n'\}$ is bounded. Choose a further subsequence $b_n'' \to b$. Then by (i) $F_n''(a_n'' x + b_n'') \xrightarrow{\mathscr{D}} \delta_{x+b}$, which implies that G is degenerate. Thus $\{a_n\}$ is indeed bounded from above. If $G_n(x) = F_n(a_n x + b_n)$ then $G_n \xrightarrow{\mathscr{D}} G$ and $G_n(a_n^{-1} x - a_n^{-1} b_n) \xrightarrow{\mathscr{D}} F$. Thus $\{a_n^{-1}\}$ is also bounded from above, and so $\{a_n\}$ must be bounded away from 0 and ∞. Finally if $\{b_n\}$ is unbounded then choose subsequences $a_n' \to a$ and $b_n' \to \pm\infty$. Since by (i) $F_n'(a_n' x) \xrightarrow{\mathscr{D}} F(ax)$ we can apply (iii) to the d.f.s $F_n'(a_n' x)$ to arrive at a contradiction.

(v) Iterating we find that $F(x) = F(a_n x + b_n)$ where $a_n = a^n$ and $b_n = (1 + a + \cdots + a^{n-1})b$. Now apply (iv). □

PROOF OF KHINCHINE'S THEOREM. By (iv) of the preceding lemma we know that $\{a_n\}$ is bounded away from 0 and ∞, and that $\{b_n\}$ is bounded. Choose a subsequence along which $a_n' \to a$ and $b_n' \to b$. Then by (i) $F_n'(a_n' x + b_n') \xrightarrow{\mathscr{D}} F(ax + b)$ and so $G(x) = F(ax + b)$. Since this uniquely determines a and b, by (v), any convergent subsequence of $\{(a_n, b_n)\}$ must converge to (a, b). □

Corollary XVII. *Let* $\mathrm{a_n, c_n} > 0$. *If* $\mathrm{F_n(a_n x + b_n)} \xrightarrow{\mathscr{D}} \mathrm{F(x)}$ *and* $\mathrm{F_n(c_n x + d_n)} \xrightarrow{\mathscr{D}} \mathrm{G(x)}$ *where* F *and* G *are nondegenerate, then* $\dfrac{\mathrm{a_n}}{\mathrm{c_n}}$ *and* $\dfrac{\mathrm{b_n - d_n}}{\mathrm{c_n}}$ *converge, and* F *and* G *are of the same type.*

PROOF. Apply Khinchine's Theorem to the d.f.s $\tilde{F}_n(x) = F_n(c_n x + d_n)$. □

A d.f. F is *extremal* if it is nondegenerate and if for some d.f. G and constants $a_n > 0$ and b_n, $G^n(a_n x + b_n) \xrightarrow{\mathscr{D}} F(x)$.

Law of Types Theorem. *Any extremal d.f. is of the same type as one of the following*: $(\alpha > 0)$

(*Type I*) $$F(x) = \exp(-e^{-x}),$$

(*Type II*) $$F(x) = \begin{cases} \exp(-x^{-\alpha}), & x \geq 0, \\ 0, & x < 0, \end{cases}$$

(*Type III*) $$F(x) = \begin{cases} 1, & x \geq 0, \\ \exp(-(-x)^{\alpha}), & x < 0. \end{cases}$$

To prove the Law of Types we need the following important tool.

Cauchy's Lemma. (i) *Let* f *be a real-valued function on* $(0, \infty)$ *satisfying*

$$f(x + y) = f(x) + f(y)$$

for all x, y > 0. *If there is some open interval on which* f *is bounded from above* (*or from below*) *then* f(x) = xf(1), x > 0.
(ii) *Let* g *be a real-valued function on* $(0, \infty)$ *satisfying*

$$g(x + y) = g(x)g(y)$$

for all x, y > 0. *If there is some interval on which* g *is bounded from above then either* g ≡ 0 *or else* $g(x) = [g(1)]^x$, x > 0.

PROOF. (i) Consider $h(x) = f(x) - xf(1)$. Then h also satisfies the functional equation $h(x + y) = h(x) + h(y)$, and $h(r) = 0$ for all $r \in \mathbb{Q}$. Suppose $h(x) \neq 0$ for some x. Since $h(r - x) = -h(x)$ for $x < r \in \mathbb{Q}$ we may assume that $h(x) > 0$. (Otherwise replace x with $r - x$.) Let I be any open interval. Given any $M > 0$ we can choose n so that $nh(x) > M$, and then choose $r \in \mathbb{Q}$ so that $nx + r \in I$. Since $h(nx + r) = nh(x) > M$ we see that h is not bounded from above on I. (Boundedness from below is treated similarly.)
(ii) Since $g(x) = g^2\left(\frac{x}{2}\right)$ we have $g \geq 0$. Observe that if g vanishes at a point it must, by the functional equation, vanish everywhere to the right of that point. If $g(x) = 0$ for some $x > 0$ then $g\left(\frac{x}{2^n}\right) = 0$ for all $n \geq 1$, and so $g \equiv 0$ in this case. Otherwise if $g(x) > 0$ for all $x > 0$ then apply (i) to $f(x) = \log g(x)$. □

PROOF OF THE LAW OF TYPES THEOREM. Assume that F is extremal. Then for some d.f. G and for each $k \geq 1$

$$G^{nk}(a_n x + b_n) \xrightarrow{\mathscr{D}} F^k(x), \qquad G^{nk}(a_{nk} x + b_{nk}) \xrightarrow{\mathscr{D}} F(x),$$

and so by Khinchine's Theorem there exist constants $c_k > 0$ and d_k such that

$$F^k(x) = F(c_k x + d_k), \qquad \forall x.$$

Since

$$F(c_{jk} x + d_{jk}) = F^{jk}(x) = F^j(c_k x + d_k) = F(c_j(c_k x + d_k) + d_j),$$

it follows from (v) of our lemma for Khinchine's Theorem that

$$c_{jk} = c_j c_k, \qquad d_{jk} = c_j d_k + d_j.$$

Of course $c_1 = 1$, $d_1 = 0$. Denote

$$x_F^+ = \inf(x\colon F(x) = 1), \qquad x_F^- = \sup(x\colon F(x) = 0).$$

We distinguish three possibilities:

(I) $c_k = 1, \forall k$;
(II) $x_F^- > -\infty$ and $d_k = x_F^-(1 - c_k), \forall k$;
(III) $x_F^+ < \infty$ and $d_k = x_F^+(1 - c_k), \forall k$.

Indeed if $c_l \neq 1$ for some $l > 1$, then we can choose a fixed point $y = c_l y + d_l$ and get $F^l(y) = F(y)$. Thus $F(y) = 0$ or 1. Suppose $F(y) = 0$. Then $x_F^- \geq y > -\infty$. If some $d_k \neq x_F^-(1 - c_k)$ then we could choose x near x_F^- so that x and $c_k x + d_k$ lie on different sides of x_F^-. But this is impossible since $F^k(x) = F(c_k x + d_k)$. Similarly if $F(y) = 1$, then $d_k = x_F^+(1 - c_k), \forall k$. We proceed now to analyze these three possibilities.

(I) In this case, $d_{jk} = d_j + d_k$ and we can consistently define

$$\eta_r = d_j - d_k \qquad \text{for} \qquad r = \frac{j}{k}.$$

Observe that $F^r(x) = F(x + \eta_r)$ for $x \in \mathbb{R}, r \in \mathbb{Q}$. Since F is nondegenerate there is an x such that $0 < F(x) < 1$. From this, it follows that d_k is strictly decreasing in k, and so η_r is strictly decreasing in r. For $t \in \mathbb{R}$ set $\varphi(t) = \inf_{\substack{0<r<t \\ r \in \mathbb{Q}}} \eta_r$. Then φ is decreasing in t, $\varphi(st) = \varphi(s) + \varphi(t)$ and

$$F^t(x) = F(x + \varphi(t)).$$

By Cauchy's Lemma $\varphi(t) = -\lambda \log t$, and so F is of Type I.

(II) We can shift so that $x_F^- = 0$. This does not change the type of F. Then $d_k = 0, \forall k$. Since $c_{jk} = c_j c_k$ we can consistently define

$$\eta_r = \frac{c_j}{c_k} \qquad \text{for} \qquad r = \frac{j}{k}.$$

Observe that $F^r(x) = F(\eta_r x)$ for $x \in \mathbb{R}, r \in \mathbb{Q}$. As in (I), since F is nondegenerate η_r is strictly decreasing in r. Define φ as earlier. Then $\varphi(st) = \varphi(s)\varphi(t)$ and

$$F^t(x) = F(\varphi(t)x).$$

By Cauchy's Lemma $\varphi(t) = t^{-\lambda}$, and so F is of Type II. The Type III analysis is similar. □

The preceding examples for the extremes show that indeed all three types do arise. A natural question to ask is characterization of the *domains of attraction*; i.e., the class of d.f.s G *attracted to* F in the sense that $G^n(a_n x + b_n) \xrightarrow{\mathscr{D}} F(x)$ for some constants $a_n > 0$ and b_n. See Leadbetter, Lindgren, and Rootzén [39] for a discussion of this.

Large Deviations (from Varadhan [57])

We are concerned here with rates of convergence for the Weak Law of Large Numbers. Let $\{X_n\}$ be an i.i.d. sequence of multivariate random variables on $\mathbb{R}^m$ with common d.f. F, finite common moment generating function $\psi(t) = \mathbf{E}\exp(\langle t, X_1\rangle)$ for all $t \in \mathbb{R}^m$, common expectation $\mu \in \mathbb{R}^m$ and partial sums $S_n = X_1 + \cdots + X_n$. The Weak Law of Large Numbers tells us that $\frac{S_n}{n} \to \mu$ in probability, and we are interested in the asymptotic decay to zero of $\mathbb{P}\left(\frac{S_n}{n} \in A\right)$, for Borel subsets $A \subseteq \mathbb{R}^m$, which do not contain μ in their closure.

Set

$$L(t) = \log \psi(t), \qquad t \in \mathbb{R}^m.$$

It follows from Hölder's inequality that L is a convex function, and it follows from Jensen's inequality that

$$L(t) \geq \langle t, \mu\rangle, \qquad t \in \mathbb{R}^m. \tag{16}$$

Furthermore, by considering the condition for equality in Hölder's inequality we find that if the X_ns are not supported inside a proper affine subset of $\mathbb{R}^m$ then L is *strictly convex*. We can make this assumption without loss of generality, since otherwise we simply work in $\mathbb{R}^{m'}$ for some $m' < m$. We consider the function

$$I(x) = \sup_{t \in \mathbb{R}^m} [\langle t, x\rangle - L(t)], x \in \mathbb{R}^m. \tag{17}$$

Being the supremum of convex functions, I itself is convex. Since $L(0) = 0$ it is clear that $I \geq 0$, and thus by (16) we infer that $I(\mu) = 0$. This function I is the *Legendre-Fenchel transform* of L. It may be infinite-valued for some points $x \in \mathbb{R}^m$. A discussion of it appears in Rockafellar [50], and there it is shown that on account of the strict convexity of L, the domain $D = \{I < \infty\}$ has nonempty interior and I is *differentiable* on D. That is, $\nabla I(x)$ exists for all points x in the interior of D, and $\lim_x \|\nabla I(x)\| = \infty$

whenever x approaches the boundary of D. See also the discussion in Ellis [15].

The supremum in (17) may or may not be attained at some $t \in \mathbb{R}^m$, and it is of interest to examine what happens when it is attained. So suppose for some $x \in \mathbb{R}^m$ there exists $t \in \mathbb{R}^m$ such that

$$I(x) = \langle t, x \rangle - L(t). \tag{18}$$

Then

$$\frac{\nabla\psi(t)}{\psi(t)} = \nabla L(t) = x. \tag{19}$$

Furthermore, it is clear in this case that the dual condition holds,

$$L(t) = \sup_{y \in \mathbb{R}^m} [\langle t, y \rangle - I(y)], \tag{20}$$

and thus

$$\nabla I(x) = t. \tag{21}$$

(In fact this dual condition (20) is always valid, even if the supremum in (17) is not attained, but we shall not be needing this fact.) We refer to I as the *rate function* for the d.f. F, on account of the following result.

Cramer's Theorem. *For any open subset* $G \subseteq \mathbb{R}^m$

$$\liminf_n \frac{1}{n} \log \mathbb{P}\left(\frac{S_n}{n} \in G\right) \geq -\inf_{x \in G} I(x), \tag{22}$$

and for any closed subset $C \subseteq \mathbb{R}^m$

$$\limsup_n \frac{1}{n} \log \mathbb{P}\left(\frac{S_n}{n} \in C\right) \leq -\inf_{x \in C} I(x). \tag{23}$$

Thus in particular if $A \subseteq \mathbb{R}^m$ *is any Borel set such that*

$$\inf_{x \in A^0} I(x) = \inf_{x \in \overline{A}} I(x),$$

then

$$\lim_n \frac{1}{n} \log \mathbb{P}\left(\frac{S_n}{n} \in A\right) = -\inf_{x \in A} I(x).$$

Proof of the Lower Bound. We first make the assumption that the smallest convex set supporting the d.f. F is all of $\mathbb{R}^m$. Then

$$\lim_{\|t\| \to \infty} \frac{L(t)}{\|t\|} = \infty. \tag{24}$$

Indeed for any $t \in \mathbb{R}^m$, $t \neq 0$, and any $N > 0$ we have $\mathbb{P}\left(\left\langle \frac{t}{\|t\|}, X_1 \right\rangle \geq N\right) \geq \alpha_N > 0$, uniformly in t, so that

$$L(t) \geq \log \alpha_N + N\|t\|.$$

On account of (24) it follows that for any $x \in \mathbb{R}^m$ the supremum in (17) must be attained at some point $t = t(x) \in \mathbb{R}^m$, and so (19) applies. Now fix a point $x \in \mathbb{R}^m$. Let us define now a new d.f. G on $\mathbb{R}^m$ by

$$G(y) = \frac{1}{\psi(t)} \int_{-\infty}^{y_1} \cdots \int_{-\infty}^{y_m} e^{\langle t, y \rangle} \, dF(y),$$

where $t = t(x)$ satisfies (18) and (19). On account of (19) the expectation for this d.f. is

$$\int_{\mathbb{R}^m} y \, dG(y) = \frac{1}{\psi(t)} \mathbf{E} X_1 e^{\langle t, X_1 \rangle} = \frac{\nabla \psi(t)}{\psi(t)} = x.$$

Thus by the Weak Law of Large Numbers, for any $\delta > 0$

$$\lim_n \psi^{-n}(t) \int \cdots \int_{B_\delta} \exp\left(\left\langle t, \sum_{i=1}^n y^{(i)} \right\rangle\right) dF(y^{(1)}) \ldots dF(y^{(n)})$$

$$= \lim_n \int \cdots \int_{B_\delta} dG(y^{(1)}) \ldots dG(y^{(n)}) = 1,$$

where B_δ is the set $\left\{ \frac{1}{n} \sum_{i=1}^n y^{(i)} \in B_\delta(x) \right\}$ and $B_\delta(x)$ is the open ball $\{\|y - x\| < \delta\}$. Thus if $\varepsilon > \delta$ then

$$\mathbb{P}\left(\frac{S_n}{n} \in B_\varepsilon(x)\right) \geq \mathbb{P}\left(\frac{S_n}{n} \in B_\delta(x)\right)$$

$$\geq \exp[-n\|t\|(\|x\| + \delta)] \int \cdots \int_{B_\delta} \exp\left(\left\langle t, \sum_{i=1}^n y^{(i)} \right\rangle\right) dF(y^{(1)}) \ldots dF(y^{(n)})$$

and thus by (18)

$$\liminf_n \frac{1}{n} \log \mathbb{P}\left(\frac{S_n}{n} \in B_\varepsilon(x)\right) \geq L(t) - \|t\|(\|x\| + \delta)$$

$$= -I(x) + \langle t, x \rangle - \|t\|(\|x\| + \delta)$$

$$\geq -I(x) - \delta\|t\|.$$

Since δ can be made arbitrarily small, we infer that for any $\varepsilon > 0$

$$\liminf_n \frac{1}{n} \log \mathbb{P}\left(\frac{S_n}{n} \in B_\varepsilon(x)\right) \geq -I(x). \tag{25}$$

The lower bound (22) follows from this since each $x \in G$ is contained in some $B_\varepsilon(x) \subseteq G$ for some $\varepsilon > 0$.

If our support assumption is not satisfied then replace X_n with $X_{n,\delta} = X_n + \delta Z_n$, where $\{Z_n\}$ is an m-dimensional i.i.d. $N(0, I_m)$ sequence, independent of the $\{X_n\}$ sequence. The moment generating function for $X_{n,\delta}$ is simply $\psi_\delta(t) = \psi(t)e^{(1/2)\delta^2\|t\|^2}$. Clearly then for $L_\delta = \log \psi_\delta$,

$$I_\delta(x) = \sup_{t \in \mathbb{R}^m} [\langle t, x \rangle - L_\delta(t)] \leq I(x).$$

Observe now that if $S_{n,\delta}$ are the partial sums $S_{n,\delta} = X_{1,\delta} + \cdots + X_{n,\delta}$ then for any $x \in \mathbb{R}^m$ and $\varepsilon > 0$

$$\mathbb{P}\left(\frac{S_n}{n} \in B_\varepsilon(x)\right) \geq \mathbb{P}\left(\frac{S_{n,\delta}}{n} \in B_{\varepsilon/2}(x)\right) - \mathbb{P}\left(\left\|\sum_{i=1}^n Z_i\right\| \geq \frac{\varepsilon}{2\delta} n\right).$$

Since $\{X_{n,\delta}\}$ does satisfy our support assumption, we can apply the lower bound (22) to estimate

$$\mathbb{P}\left(\frac{S_{n,\delta}}{n} \in B_{\varepsilon/2}(x)\right) \geq \exp[-nI_\delta(x) + o(n)]$$

as $n \to \infty$. On the other hand, since $\dfrac{1}{n}\left\|\sum_{i=1}^n Z_i\right\|^2$ is distributed according to χ^2 with m degrees of freedom

$$\mathbb{P}\left(\left\|\sum_{i=1}^n Z_i\right\| \geq \frac{\varepsilon}{2\delta} n\right) \leq \exp\left[-\frac{\varepsilon^2}{8\delta^2} n + o(n)\right]$$

as $n \to \infty$. Choosing δ so small now that $\dfrac{\varepsilon^2}{8\delta^2} \geq I(x)$ and combining these estimates leads to (25). □

Proof of the Upper Bound. For any Borel subset $C \subseteq \mathbb{R}^m$ and for any $t \in \mathbb{R}^m$

$$\mathbb{P}\left(\frac{S_n}{n} \in C\right) \leq \mathbf{E} \exp\left(\left\langle t, \frac{S_n}{n}\right\rangle - \inf_{x \in C} \langle t, x \rangle\right),$$

so that

$$\frac{1}{n} \log \mathbb{P}\left(\frac{S_n}{n} \in C\right) \leq -\frac{1}{n} \inf_{x \in C} \langle t, x \rangle + L\left(\frac{t}{n}\right).$$

Replacing t by nt, and then taking the supremum over all $t \in \mathbb{R}^m$ leads to

$$\frac{1}{n} \log \mathbb{P}\left(\frac{S_n}{n} \in C\right) \leq -\sup_t \inf_{x \in C} [\langle t, x \rangle - L(t)].$$

We now invoke the Minimax Theorem (proved later), which asserts that *for any compact convex subset* $D \subseteq \mathbb{R}^m$

$$\inf_{x \in C} I(x) = \inf_{x \in C} \sup_t [\langle t, x \rangle - L(t)] = \sup_t \inf_{x \in C} [\langle t, x \rangle - L(t)]. \quad (26)$$

This establishes (23) for compact convex subsets C.

Next let K be any compact subset of $\mathbb{R}^m$, and let l be such that $I(x) \geq l$ for all $x \in K$. Given $\varepsilon > 0$ it follows from the continuity of I that for each $x \in K$ we can find a closed ball $C(x)$ of some finite positive radius such that $I(y) \geq l - \varepsilon$ for all $y \in C(x)$. (Even though I may be infinite-valued, it is always lower semicontinuous – the supremum of continuous functions; thus sets of the form $\{I > \alpha\}$ are always open.) Let $C(x_1), \ldots, C(x_N)$ be a finite subcover of K extracted from these. Then

$$\mathbb{P}\left(\frac{S_n}{n} \in K\right) \leq \sum_{i=1}^{N} \mathbb{P}\left(\frac{S_n}{n} \in C(x_i)\right) \leq N \max_{1 \leq i \leq N} \mathbb{P}\left(\frac{S_n}{n} \in C(x_i)\right).$$

Since each set $C(x)$ is compact and convex we can apply the upper bound (23) to infer that

$$\mathbb{P}\left(\frac{S_n}{n} \in K\right) \leq N \exp[-n(l - \varepsilon) + o(n)]$$

as $n \to \infty$. Since ε was arbitrary, we conclude that (23) holds for compact subsets $C = K$ as well.

Suppose for the moment that $m = 1$. By Chebyshev's inequality, for any $\rho > 0$ and any n

$$\mathbb{P}\left(\left|\frac{S_n}{n}\right| > \rho\right) \leq \{e^{-\rho}[\psi(1) + \psi(-1)]\}^n.$$

Thus given any $\lambda < \infty$ we can always choose ρ large enough so that $\frac{1}{n} \log \mathbb{P}\left(\left|\frac{S_n}{n}\right| > \rho\right) \leq -\lambda$ for every n. Back to the m-dimensional setting, then, applying this to the ith components of the X_ns: given any $\lambda > 0$ we can always choose ρ large enough so that for any n

$$\frac{1}{n} \log \mathbb{P}\left(\frac{S_n}{n} \in A_{\rho,i}^c\right) \leq -\lambda,$$

where $A_{\rho,i} = \{x \in \mathbb{R}^m : |x_i| \leq \rho\}$. Thus if K_ρ is the compact set $K_\rho = \bigcap_{i=1}^{m} A_{\rho,i}$ then

$$\limsup_n \frac{1}{n} \log \mathbb{P}\left(\frac{S_n}{n} \in K_\rho^c\right) \leq -\lambda. \quad (27)$$

Now let C be any closed subset of $\mathbb{R}^m$, and suppose $I(x) \geq l$ for all $x \in C$. Then for any $\rho > 0$

$$\mathbb{P}\left(\frac{S_n}{n} \in C\right) \leq \mathbb{P}\left(\frac{S_n}{n} \in C \cap K_\rho\right) + \mathbb{P}\left(\frac{S_n}{n} \in K_\rho^c\right).$$

Choose now $\lambda \leq l$ and correspondingly choose ρ so that (27) holds. Then by applying our upper bound (23) to the compact set $C \cap K_\rho$ we estimate

$$\limsup_n \frac{1}{n} \log \mathbb{P}\left(\frac{S_n}{n} \in C\right) \leq \max(-l, -\lambda) = -l.$$

Thus (23) holds for this general closed set C. □

There remains to discuss the Minimax Theorem (26). The left-hand side of this equality is always greater than or equal to the right-hand side (for any set C), so we can concern ourselves with the opposite inequality. Furthermore, if $\mu \in C$ then the left-hand side is zero, so we assume in addition that $\mu \notin C$. Suppose first, as we did in the proof of the lower bound, that the support of the X_ns is not contained in any proper convex subset of $\mathbb{R}^m$; so that the supremum in (17) is always attained. Since C is compact the infimum $\inf_{x \in C} I(x)$ is attained, say, at the point $y \in C$. Since C is convex the tangent hyperplane for the level surface of I at y must be supporting for C. Thus

$$\inf_{x \in C} \langle x, \nabla I(y) \rangle = \langle y, \nabla I(y) \rangle.$$

Using (18) and (21) leads now to (26).

For the general case we perturb I by

$$I_\delta(x) = \sup_t \left[\langle x, t \rangle - L(t) - \delta \|t\|^2\right] \tag{28}$$

for $\delta > 0$, just as we did in the proof of the lower bound. The supremum in (28) is always attained, and so (26) holds for each rate function I_δ. Thus to prove (26) for I it suffices to establish that

$$\inf_{x \in C} I_\delta(x) \uparrow \inf_{x \in C} I(x) \quad \text{as} \quad \delta \downarrow 0. \tag{29}$$

For each $\delta > 0$ let $I_\delta(x_\delta) = \inf_{x \in C} I_\delta(x)$, where $x_\delta \in C$. Since C is compact some subsequence $\{x_{\delta_n}\}$ converges to a limit $y \in C$, for $\delta_n \to 0$. Suppose $I(y) < \infty$. Then given any $\varepsilon > 0$ we can choose $t = t(\varepsilon)$ such that

$$\langle y, t \rangle - L(t) \geq I(y) - \varepsilon.$$

Since

$$I_\delta(x_\delta) \geq \langle x_\delta, t \rangle - L(t) - \delta \|t\|^2$$

we have

$$\lim_\delta I_\delta(x_\delta) = \lim_n I_{\delta_n}(x_{\delta_n}) \geq \langle y, t \rangle - L(t)$$

$$\geq I(y) - \varepsilon \geq \inf_{x \in C} I(x) - \varepsilon.$$

Since $\varepsilon > 0$ was arbitrary we obtain (29). In case $I(y) = \infty$ then replace $I(y) - \varepsilon$ with an arbitrary M in the preceding argument. □

Examples

Suppose each X_n is Bernoulli with $p = \frac{1}{2}$. Then $\psi(t) = \frac{1}{2}(1 + e^t)$ and we compute

$$I(x) = \begin{cases} \log 2 + x \log x + (1 - x)\log(1 - x), & 0 \le x \le 1, \\ \infty, & \text{otherwise.} \end{cases}$$

Since S_n is binomial

$$\mathbb{P}\left(\frac{S_n}{n} = \frac{j}{n}\right) = \frac{\binom{n}{j}}{2^n},$$

and so by Stirling's approximation for large j, n

$$\frac{1}{n}\log \mathbb{P}\left(\frac{S_n}{n} = \frac{j}{n}\right) \sim -I\left(\frac{j}{n}\right).$$

Suppose next that each X_n is exponential with mean one. Then $\psi(t) = \dfrac{1}{1-t}$ for $t < 1$, and we compute

$$I(x) = \begin{cases} x - 1 - \log x, & x > 0, \\ \infty, & x \le 0. \end{cases}$$

Since S_n is gamma-distributed we can find the density of $\dfrac{S_n}{n}$ explicitly, and it turns out to be

$$f_n(x) = \frac{n^n}{(n-1)!}x^{n-1}e^{-nx}, \qquad x > 0;$$

again by Stirling's approximation we verify that

$$\frac{1}{n}\log f_n(x) \sim -I(x).$$

(N.B. Strictly speaking, in this example we are using an extension of Cramér's Theorem, since $\psi(t)$ is only defined for $t < 1$.)

Suppose next that each X_n is m-dimensional $N(\mu, \Gamma)$, where the covariance matrix Γ is nonsingular. Then

$$\psi(t) = \exp(\langle t, \mu\rangle + \tfrac{1}{2}\langle \Gamma t, t\rangle),$$

and by minimizing a positive definite quadratic form, the rate function works out to be

$$I(x) = \tfrac{1}{2}\langle \Gamma^{-1}(x - \mu), x - \mu\rangle.$$

Since $\frac{S_n}{n}$ is $N\left(\mu, \frac{1}{n}\Gamma\right)$ its density is given by

$$f_n(x) = \frac{n^{m/2}}{(2\pi)^{m/2}\sqrt{\det \Gamma}} \exp\left[-\frac{n}{2}\langle \Gamma^{-1}(x-\mu), x-\mu\rangle\right];$$

and manifestly

$$\frac{1}{n}\log f_n(x) \sim -I(x).$$

Suppose each X_n has the atomic distribution on $\mathbb{R}^m$

$$\mathbb{P}(X_n = e^{(i)}) = p_i, \qquad 1 \le i \le m;$$

where $e^{(1)}, \ldots, e^{(m)}$ are the unit vectors along the axes, and p_i are positive numbers summing to one. The moment generating function is

$$\psi(t) = \sum_{i=1}^{m} p_i e^{t_i},$$

and the rate function works out to be

$$I(x) = \begin{cases} \sum_{i=1}^{m} x_i \log \frac{x_i}{p_i}, \text{ if } x_i \ge 0 \text{ and } \sum_{i=1}^{m} x_i = 1, \\ \infty, \text{ otherwise.} \end{cases}$$

Observe that $\frac{S_n}{n}$ has the multinomial distribution

$$\mathbb{P}\left(\frac{S_n}{n} = \left(\frac{j_1}{n}, \ldots, \frac{j_m}{n}\right)\right) = \frac{n!}{j_1! \ldots j_m!} p_1^{j_1} \ldots p_m^{j_m},$$

whenever the j_is are nonnegative integers summing to n. As in the binomial example, we check, using Stirling's approximation, that for large n, $j_1, \ldots, j_m$

$$\frac{1}{n}\log \mathbb{P}\left(\frac{S_n}{n} = \left(\frac{j_1}{n}, \ldots, \frac{j_m}{n}\right)\right) \sim -I\left(\frac{j_1}{n}, \ldots, \frac{j_m}{n}\right).$$

The rate function here is the relative entropy, or the Kullback-Liebler information of the discrete probability density (x_i) relative to (p_i).

Exercises

1. (a) Let X be distributed according to the Poisson law, $\mathbf{E}X = \lambda$. Prove that as $\lambda \to \infty$

$$\frac{X - \lambda}{\sqrt{\lambda}}$$

tends weakly to $N(0, 1)$.

(b) Let X be gamma-distributed with parameters r and λ. Prove that as $r \to \infty$

$$\frac{\lambda X - r}{\sqrt{r}}$$

tends weakly to $N(0, 1)$.

2. (Gnedenko [23]) Prove that as $n \to \infty$

$$e^{-n} \sum_{k=0}^{n} \frac{n^k}{k!} \to 1/2.$$

(Hint: Consider sums of independent Poissons, mean one.)

3. (Gnedenko [23]) In 14,400 tosses of a coin, heads had come up 7,428 times. What is the probability of the number of heads deviating from the quantity np by an amount equal to or greater than the deviation in this experiment, if the coin is symmetric (i.e., the probability of tossing a head in each trial is 1/2)?

4. (Gnedenko [23]) 10,000 persons of the same age and social group have policies with an insurance company. The probability of death during the year is 0.006 for each one of them. Each insured pays a premium of \$12 on January 1, and in the event he dies, his beneficiaries receive \$1000 from the company. What is the probability that

(a) the company will lose money;
(b) the company will make a profit of not less than \$40,000? \$60,000? \$80,000?

Section IV
Markov Chains—Passage Phenomena*

My treatment of Markov chains in the three chapters which follow is modelled after the material in "Introduction to Stochastic Processes" by Hoel, Port and Stone (Ref. [28]), and is presented here with their kind permission. I have adopted their notation and style, because I feel it is the best way to introduce Markov chains in the spirit of these notes—namely, an approach which combines intuition (of the dynamics) with probabilistic reasoning. The presentation here is compressed and condensed. For a more leisurely account of this material, replete with many examples, problems and related topics, I recommend the Hoel, Port and Stone text. In addition their text discusses the important topics of differentiation and integration of stochastic processes, Brownian motion and stochastic differential equations, which are not contained herein.

First Notions and Results

Definition I. Let $\{X_n\}_{n\geq 0}$ be a sequence of random variables taking values in a finite or countably infinite set $\mathscr{L}$. It is said to be a *Markov chain* if

$$\mathbb{P}(X_{n+1} = x_{n+1} | X_0 = x_0, \ldots, X_n = x_n) = \mathbb{P}(X_{n+1} = x_{n+1} | X_n = x_n) \quad (1)$$

for every choice of the nonnegative number n and $x_0, \ldots, x_{n+1} \in \mathscr{L}$. We will always assume that $\mathscr{L}$ is a subset of the integers. The conditional probabilities $\mathbb{P}(X_{n+1} = y | X_n = x)$ are called the *transition probabilities* of the chain. The Markov chain is said to be *stationary* if the transition probabilities are independent of n:

$$\mathbb{P}(X_{n+1} = y | X_n = x) = P(x, y). \quad (2)$$

* Material taken from Hoel, Port and Stone [28] with permission.

The function $P(x, y)$, $x, y \in \mathscr{L}$ satisfies

$$P(x, y) \geq 0, \qquad x, y \in \mathscr{L}, \tag{3}$$

$$\sum_y P(x, y) = 1, \qquad x \in \mathscr{L}. \tag{4}$$

If $\mathscr{L}$ is finite, say $\mathscr{L} = \{0, 1, \ldots, d\}$, we can represent P as a $(d + 1) \times (d + 1)$ matrix

$$\begin{bmatrix} P(0, 0) \ldots P(0, d) \\ \vdots \qquad \vdots \\ P(d, 0) \ldots P(d, d) \end{bmatrix}.$$

The probability π_0 defined by

$$\pi_0(x) = \mathbb{P}(X_0 = x) \tag{5}$$

is called the *initial distribution* of the chain. We shall use the notation $\mathbb{P}_x(A)$ to denote $\mathbb{P}(A | X_0 = x)$, and $\mathbf{E}_x Y$ to denote $\mathbf{E}(Y | X_0 = x)$.

Theorem I. (i) *For* $x_0, \ldots, x_n \in \mathscr{L}$

$$\mathbb{P}(X_0 = x_0, \ldots, X_n = x_n) = \pi_0(x_0) P(x_0, x_1) \ldots P(x_{n-1}, x_n).$$

(ii) *For subsets* $A_0, \ldots, A_{n-1}, B_1, \ldots, B_m \subseteq \mathscr{L}$ *and* $x \in \mathscr{L}$

$$\mathbb{P}(X_{n+1} \in B_1, \ldots, X_{n+m} \in B_m | X_0 \in A_0, \ldots, X_{n-1} \in A_{n-1}, X_n = x)$$
$$= \sum_{y_1 \in B_1} \cdots \sum_{y_m \in B_m} P(x, y_1) P(y_1, y_2) \ldots P(y_{m-1}, y_m).$$

(iii) *For* $x, y \in \mathscr{L}$

$$\mathbb{P}(X_n = y | X_0 = x) = P^n(x, y),$$

where P^n *denotes the* nth *power of the (finite or countably infinite) matrix* P. *If* $\pi_n(x) = \mathbb{P}(X_n = x)$ *then*

$$\pi_n = \pi_0 P^n,$$

where π_n *is the (finite or countably infinite) row vector* $(\pi_n(x))_{x \in \mathscr{L}}$.

PROOF. (i)

$$\mathbb{P}(X_0 = x_0, \ldots, X_n = x_n)$$
$$= \mathbb{P}(X_n = x_n | X_0 = x_0, \ldots, X_{n-1} = x_{n-1}) \mathbb{P}(X_0 = x_0, \ldots, X_{n-1} = x_{n-1})$$
$$= P(x_{n-1}, x_n) \mathbb{P}(X_0 = x_0, \ldots, X_{n-1} = x_{n-1}).$$

Now use induction on n. □

(ii) $\mathbb{P}(X_0 \in A_0, \ldots, X_{n-1} \in A_{n-1}, X_n = x, X_{n+1} \in B_1, \ldots, X_{n+m} \in B_m)$

$$= \sum_{x_0 \in A_0} \cdots \sum_{x_{n-1} \in A_{n-1}} \sum_{y_1 \in B_1} \cdots \sum_{y_m \in B_m} \pi_0(x_0) P(x_0, x_1) \ldots P(x_{n-1}, x)$$
$$\cdot P(x, y_1) \ldots P(y_{m-1}, y_m)$$

$$= \left[\sum_{x_0 \in A_0} \cdots \sum_{x_{n-1} \in A_{n-1}} \pi_0(x_0) P(x_0, x_1) \ldots P(x_{n-1}, x)\right]$$

$$\cdot \left[\sum_{y_1 \in B_1} \cdots \sum_{y_m \in B_m} P(x, y_1) \ldots P(y_{m-1}, y_m)\right]$$

$$= \mathbb{P}(X_0 \in A_0, \ldots, X_{n-1} \in A_{n-1}, X_n = x)$$

$$\cdot \left[\sum_{y_1 \in B_1} \cdots \sum_{y_m \in B_m} P(x, y_1) \ldots P(y_{m-1}, y_m)\right]. \qquad \square$$

(iii) $$\pi_n(y) = \mathbb{P}(X_n = y) = \sum_x \mathbb{P}(X_n = y | X_{n-1} = x) \mathbb{P}(X_{n-1} = x)$$

$$= \sum_x \pi_{n-1}(x) P(x, y).$$

Thus by induction on n, $\pi_n = \pi_0 P^n$. Setting $\pi_0 = \delta_x$ we see that

$$\mathbb{P}_x(X_n = y) = P^n(x, y). \qquad \square$$

Definition II. Let A be a subset of $\mathscr{L}$. The *hitting time* T_A is defined by

$$T_A = \begin{cases} \min(n > 0 : X_n \in A), & \text{if} \quad X_n \in A \text{ for some } n > 0, \\ \infty, & \text{otherwise.} \end{cases}$$

In other words T_A is the first positive time the Markov chain is in (i.e., hits) the set A. The hitting time of a singleton $\{a\}$ is denoted T_a. For $x, y \in \mathscr{L}$ set

$$\rho_{xy} = \mathbb{P}_x(T_y < \infty). \tag{6}$$

If $\rho_{xy} > 0$ we say that x *leads to* y. A nonempty set $C \subseteq \mathscr{L}$ is said to be *closed* if no state inside of C leads to any state outside of C; that is, if

$$\rho_{xy} = 0, \quad x \in C, \quad y \notin C. \tag{7}$$

It is said to be *irreducible* if x leads to y for all choices $x, y \in C$. A state y is called *recurrent* if $\rho_{yy} = 1$ and *transient* if $\rho_{yy} < 1$. If all states $y \in \mathscr{L}$ are recurrent (resp., transient) the chain is called a *recurrent* (resp. *transient*) *chain.*

Theorem II. *Let* N(y) *denote the number of times* $n \geq 1$ *that the chain is in state* y.

(i) x *leads to* y *if and only if* $\mathbf{E}_x N(y) > 0$.
(ii) *If* y *is recurrent then*

$$\mathbb{P}_x(N(y) = \infty) = \rho_{xy}, \qquad x \in \mathscr{L}. \tag{8}$$

(iii) *If* y *is transient then*

$$\mathbf{E}_x N(y) = \frac{\rho_{xy}}{1 - \rho_{yy}}, \qquad x \in \mathscr{L}. \tag{9}$$

In particular

$$\sum_{n=1}^{\infty} P^n(x, y) < \infty, \qquad x \in \mathscr{L}, \tag{10}$$

and any finite closed set must have a recurrent state.

PROOF. (i) We first observe that x leads to y if and only if $P^n(x, y) > 0$ for some positive integer n. This is because

$$\{T_y = n\} \subseteq \{X_n = y\} \subseteq \{T_y \le n\}.$$

Next, observe that

$$N(y) = \sum_{n=1}^{\infty} I_{\{y\}}(X_n),$$

and thus

$$\mathbf{E}_x N(y) = \sum_{n=1}^{\infty} \mathbb{P}_x(X_n = y) = \sum_{n=1}^{\infty} P^n(x, y). \qquad \square \tag{11}$$

(ii) Observe first that

$$\mathbb{P}_x(N(y) \ge m) = \rho_{xy}\rho_{yy}^{m-1}, \qquad m \ge 1. \tag{12}$$

This is because

$$\begin{aligned}\mathbb{P}_x(N(y) \ge m) &= \mathbb{P}_x(N(y) \ge m | N(y) \ge m-1)\mathbb{P}_x(N(y) \ge m-1) \\ &= \mathbb{P}_y(N(y) \ge 1)\mathbb{P}_x(N(y) \ge m-1) = \rho_{yy}\mathbb{P}_x(N(y) \ge m-1),\end{aligned}$$

and so (12) follows inductively on m. Since $\rho_{yy} = 1$ and

$$\mathbb{P}_x(N(y) = \infty) = \lim_m \mathbb{P}_x(N(y) \ge m)$$

the conclusion (8) follows. $\square$

(iii) $$\mathbf{E}_x N(y) = \sum_{m=1}^{\infty} \mathbb{P}_x(N(y) \ge m) = \sum_{m=1}^{\infty} \rho_{xy}\rho_{yy}^{m-1} = \frac{\rho_{xy}}{1 - \rho_{yy}}.$$

The summability condition (10) follows next by (11). If C is a finite closed set of states it must contain a recurrent state; otherwise if all states were transient then

$$\sum_{y \in C} \sum_{n=1}^{\infty} P^n(x, y)$$

would be finite (for any choice of $x \in C$), whereas

$$\sum_{n=1}^{\infty} \sum_{y \in C} P^n(x, y) = \sum_{n=1}^{\infty} 1 = \infty. \qquad \square$$

Let $\mathscr{L}_R$ and $\mathscr{L}_T$ denote the sets of recurrent and transient points of $\mathscr{L}$, respectively.

Theorem III. *Assume $\mathscr{L}_R \neq \phi$. Then $\mathscr{L}_R$ is closed and the relation "leads to" is an equivalence relation on $\mathscr{L}_R$. Accordingly $\mathscr{L}_R$ partitions into irreducible closed sets $C_1, C_2, \ldots$. Furthermore, if x is recurrent then for any y, $\rho_{xy} = 0$ or 1.*

Proof. Suppose $\rho_{xy} > 0$ and let k be the smallest positive integer for which $P^k(x, y) > 0$. Since no path of positive probability that goes from x to y in k steps can have x as an intermediate point (on account of the minimality of k), it follows that

$$\mathbb{P}_x(T_x = \infty) \geq P^k(x, y)\mathbb{P}_y(T_x = \infty).$$

Thus if $\rho_{xx} = 1$ then $\rho_{yx} = 1$. In particular since $\rho_{yx} > 0$ we can choose l such that $P^l(y, x) > 0$. Then for any $n \geq k + l$

$$P^n(y, y) \geq P^l(y, x)P^{n-k-l}(x, x)P^k(x, y).$$

Since $\rho_{xx} = 1$ it follows from our previous theorem that $\sum_{n=k+l}^{\infty} P^n(y, y) = \infty$, and thus $\rho_{yy} = 1$. We have thus established that if x is recurrent and x leads to y then y is recurrent and $\rho_{yx} = 1$. Finally, for any $k, l \geq 0$

$$P^{k+l}(x, z) \geq P^k(x, y)P^l(y, z),$$

and so "leads to" is a transitive relation. □

We address next the problem of computing ρ_{xy} when $x \in \mathscr{L}_T$, $y \in \mathscr{L}_R$. In this case y lies in a closed irreducible set C, and $\rho_{xy} = \rho_C(x) = \mathbb{P}_x(T_C < \infty)$ is the probability of absorption. We can set up an implicit system of linear equations for these probabilities. Observe that if $x \in \mathscr{L}_T$, a chain starting at x can enter C only by entering C at time 1 or by being in $\mathscr{L}_T$ at time 1 and entering C at some future time. Thus

$$\rho_C(x) = \sum_{y \in C} P(x, y) + \sum_{y \in \mathscr{L}_T} P(x, y)\rho_C(y), \qquad x \in \mathscr{L}_T.$$

We need to concern ourselves with the uniqueness of solutions to this linear system.

Theorem IV (from Parzen [45]). *The homogeneous system of equations*

$$f(x) = \sum_{y \in \mathscr{L}_T} P(x, y)f(y), \qquad x \in \mathscr{L}_T, \tag{13}$$

has a unique bounded solution $\{f(x): x \in \mathscr{L}_T\}$ (namely, the zero solution) if and only if

$$\mathbb{P}_x(T_{\mathscr{L}_R} < \infty) = 1, \qquad x \in \mathscr{L}_T. \tag{14}$$

Proof. Let $f(x) = \mathbb{P}_x(T_{\mathscr{L}_R} = \infty)$, $x \in \mathscr{L}_T$. Then f satisfies (13). Thus if (13) has only the trivial bounded solution, then (14) follows. Conversely assume (14) holds, and let $g(x)$ be any bounded solution of (13). Assume without loss of generality that $|g| \le 1$. (Otherwise just normalize it.) Now

$$\begin{aligned}\mathbb{P}_x(T_{\mathscr{L}_R} > 1) &= \mathbb{P}_x(X_1 \in \mathscr{L}_T) = \sum_{y \in \mathscr{L}_T} P(x, y) \\ &\ge \left| \sum_{y \in \mathscr{L}_T} P(x, y) g(y) \right| = |g(x)|.\end{aligned}$$

Similarly

$$\begin{aligned}\mathbb{P}_x(T_{\mathscr{L}_R} > 2) &= \mathbb{P}_x(X_2 \in \mathscr{L}_T) = \sum_{y \in \mathscr{L}_T} P(x, y) \mathbb{P}_y(T_{\mathscr{L}_R} > 1) \\ &\ge \sum_{y \in \mathscr{L}_T} P(x, y)|g(y)| \ge \left| \sum_{y \in \mathscr{L}_T} P(x, y) g(y) \right| = |g(x)|,\end{aligned}$$

and thus inductively

$$|g(x)| \le \mathbb{P}_x(T_{\mathscr{L}_R} > n).$$

Letting $n \to \infty$ we obtain

$$|g(x)| \le \mathbb{P}_x(T_{\mathscr{L}_R} = \infty) = 0. \qquad \square$$

If $\mathscr{L}_T$ is finite then (14) automatically holds, since a transient state is only visited a finite number of times.

If $x, y \in \mathscr{L}_T$ we have

$$\rho_{xy} = P(x, y) + \sum_{\substack{z \in \mathscr{L}_T \\ z \ne y}} P(x, z) \rho_{zy}, \qquad x \in \mathscr{L}_T.$$

For each fixed $y \in \mathscr{L}_T$ we have here a linear system for $(\rho_{xy})_{\substack{x \in \mathscr{L}_T \\ x \ne y}}$. The necessary and sufficient condition for a unique solution here is that

$$\mathbb{P}_x(T_{\mathscr{L}_R \cup \{y\}} < \infty) = 1, \qquad x \in \mathscr{L}_T, x \ne y.$$

To see this, replace P with $\hat{P}$ defined by

$$\hat{P}(x, z) = \begin{cases} P(x, z), & x \ne y \\ 1, & x = y = z \\ 0, & x = y, z \ne y. \end{cases}$$

This turns y into an absorbing state, so that $\hat{\mathscr{L}}_R = \mathscr{L}_R \cup \{y\}$, where $\hat{\mathscr{L}}_R$ is the set of recurrent states under the transition structure $\hat{P}$. Now fall back on our previous theorem.

We turn next to the mean first passage times

$$m_{xy} = \mathbf{E}_x T_y.$$

Of course if $\rho_{xy} < 1$, then $m_{xy} = \infty$, so we concern ourselves with an irreducible recurrent Markov chain. In this case we arrive at the linear system

$$m_{xy} = 1 + \sum_{z \ne y} P(x, z) m_{zy}.$$

As earlier, for each fixed y this is a linear system for $(m_{xy})_{x \neq y}$. By considering $\hat{P}$ again we can convince ourselves that this system always has a unique *bounded* solution, but it may happen that the m_{xy} are unbounded or even infinite when $\mathscr{L}$ is infinite. We shall elaborate on this when we discuss the notion of positive recurrence in the next section. Note that under $\hat{P}$, since $\mathscr{L}_R = \{y\}$,

$$\hat{m}_{xy} = 1 + \sum_{z \neq y} \hat{\mathbf{E}}_x N(z).$$

On the other hand since m_{xy} is a *first* passage time, $m_{xy} = \hat{m}_{xy}$. Thus by (11) we infer that

$$m_{xy} = \sum_{z \neq y} \sum_{n=0}^{\infty} \hat{P}^n(x, z).$$

A birth and death chain is a Markov chain on $\{0, 1, \ldots, d\}$ or $\{0, 1, 2, \ldots\}$ with transition probabilities satisfying

$$P(x, y) = 0 \quad \text{if} \quad |x - y| \geq 2.$$

Then let us denote

$$P(x, y) = \begin{cases} q_x, & y = x - 1 \\ r_x, & y = x \\ p_x, & y = x + 1, \end{cases}$$

where $p_x, q_x, r_x \geq 0$ satisfy $p_x + q_x + r_x = 1$.

Theorem V. *Let* $\{X_n\}_{n \geq 0}$ *be a birth and death chain on* $\{0, 1, \ldots, d\}$. *Assume* $p_x > 0$, $0 < x < d$. *Define*

$$\gamma_y = \begin{cases} \dfrac{q_1 \cdots q_y}{p_1 \cdots p_y}, & y \geq 1, \\ 1, & y = 0. \end{cases}$$

For states a *and* b *with* $a < b$,

$$\mathbb{P}_x(T_a < T_b) = \frac{\sum_{y=x}^{b-1} \gamma_y}{\sum_{y=a}^{b-1} \gamma_y}, \qquad a < x < b. \tag{15}$$

In particular if 0, d *are absorbing states, then*

$$\rho_{\{0\}}(x) = \frac{\sum_{y=x}^{d-1} \gamma_y}{\sum_{y=0}^{d-1} \gamma_y}, \qquad x > 0.$$

Proof. Let us denote by L the operator

$$(Lu)(x) = q_x u(x-1) - (q_x + p_x)u(x) + p_x u(x+1),$$

where u is a sequence. Observe that if u is chosen as $u(x) = \mathbb{P}_x(T_a < T_b)$, $a < x < b$, then it satisfies the difference equation

$$Lu(x) = 0, \qquad a < x < b, \tag{16}$$

with boundary conditions

$$u(a) = 1, \qquad u(b) = 0. \tag{17}$$

By identifying

$$\begin{aligned} u(x+1) - u(x) &= \frac{q_x}{p_x}[u(x) - u(x-1)] \\ &= \cdots = \frac{\gamma_x}{\gamma_a}[u(a+1) - u(a)], \end{aligned}$$

one can solve (16) and (17) to arrive at (15). □

Corollary VI. *Let* $\{X_n\}_{n\geq 0}$ *be a birth and death chain on* $\{0, 1, \dots\}$. *Assume* $p_x > 0, x > 0$. *For any state* a

$$\mathbb{P}_x(T_a < \infty) = 1 - \frac{\sum_{y=a}^{x-1} \gamma_y}{\sum_{y=a}^{\infty} \gamma_y}, \qquad x > a. \tag{18}$$

In particular if the chain is irreducible then it is transient if and only if $\sum_y \gamma_y < \infty$.

PROOF. Observe that for $n > a$

$$\{T_a < T_n\} \subseteq \{T_a < T_{n+1}\}$$

and $T_n \to \infty$ a.s. Thus

$$\mathbb{P}_x(T_a < \infty) = \lim_n \mathbb{P}_x(T_a < T_n).$$

From this we arrive at (18), and there the sufficiency of $\sum_y \gamma_y < \infty$ follows for transience. Conversely, since $\rho_{00} = r_0 + p_0\rho_{10}$ it follows that this condition is also necessary. □

Similarly let us try to compute the mean first passage times. Suppose the chain is irreducible, and set $v(x) = \mathbf{E}_x(T_a \wedge T_b)$, $a < x < b$, and take $v(a) = v(b) = 0$ for consistency in what follows. Then v satisfies the difference equation

$$Lv(x) = -1, \qquad a < x < b,$$

with boundary conditions

$$v(a) = v(b) = 0.$$

Based on the fact that

$$v(x+1) - v(x) = \frac{\gamma_x}{\gamma_a}[v(a+1) - v(a)] - \sum_{y=a+1}^{x} \frac{\gamma_x}{p_y \gamma_y},$$

for $a < x < b$, we calculate

$$v(x) = \frac{\sum_{y=a}^{x-1} \gamma_y}{\sum_{y=a}^{b-1} \gamma_y} \sum_{y=a}^{b-1} \sum_{z=a}^{y} \frac{\gamma_y}{p_z \gamma_z} - \sum_{y=a}^{x-1} \sum_{z=a}^{y} \frac{\gamma_y}{p_z \gamma_z}, \qquad a < x < b.$$

In particular if $\mathscr{L} = \{0, 1, 2, \ldots\}$ and $\rho_{xa} = 1$ then

$$\mathbf{E}_x T_a = \lim_n \mathbf{E}_x(T_a \wedge T_n) = \sum_{y=a}^{x-1} \sum_{z=y+1}^{\infty} \frac{\gamma_y}{p_z \gamma_z},$$

for $x > a$.

Next set $w(x) = \mathbf{E}_x T_b$, $x < b$, and take $w(b) = 0$ for consistency in what follows. Then w satisfies the difference equation

$$Lw(x) = -1, \qquad 0 < x < b,$$

with boundary conditions

$$w(0) - w(1) = \frac{1}{p_0}, \qquad w(b) = 0.$$

We solve this to arrive at

$$w(x) = \sum_{y=x}^{b-1} \sum_{z=0}^{y} \frac{\gamma_y}{p_z \gamma_z}, \qquad x \le b.$$

Finally, now, if $\mathscr{L} = \{0, 1, 2, \ldots\}$ then for $x \ge 0$,

$$\mathbf{E}_x T_x = 1 + q_x \mathbf{E}_{x-1} T_x + p_x \mathbf{E}_{x+1} T_x = p_x \gamma_x \sum_{z=0}^{\infty} \frac{1}{p_z \gamma_z}.$$

Other than the case of an i.i.d. sequence $\{X_n\}$, whereby $P(x, y)$ is independent of x altogether, the most primitive example of a Markov chain is the sums of i.i.d.s; i.e. $X_n = \sum_{k=0}^{n} Z_k$, where $\{Z_n\}$ is an i.i.d. integer-valued sequence. Such a chain is called a *random walk*, and for it there holds $P(x, y) = p_Z(y - x)$, where p_Z is the common discrete density for the Z_ns. This chain will be a birth and death chain whenever the Z_ns take only the values 0, ± 1; and in this case we refer to the chain as *simple random walk*. Typically the state space for simple random walk is the entire set of integers $\mathscr{L} = \{0, \pm 1, \pm 2, \ldots\}$. If we want to confine it to $\mathscr{L} = \{0, 1, \ldots, d\}$ or to $\mathscr{L} = \{0, 1, 2, \ldots\}$ we need to impose boundary conditions to ensure that it does not "walk through" 0 and/or d. These confined chains are sometimes referred to as "gambler's ruin," based on the interpretation of the Z_ns as the winning/loss of a gamble. Here $p = \mathbb{P}(Z_n = 1)$ could represent the probability of winning one dollar, and $q = 1 - p = \mathbb{P}(Z_n = -1)$ could represent the probability of losing one dollar. When two gamblers oppose each other the state space is $\{0, 1, \ldots, d\}$, where d is the sum of their stakes, and 0 and d are absorbing states (one player goes broke). When one gambler plays the house (considered infinitely wealthy) the state space is $\{0, 1, 2, \ldots\}$, and 0 is an absorbing state. Our calculations for general birth and death chains apply to simple random walks.

Let us consider, for example, the chain on $\{0, 1, 2, \ldots\}$ with transition probabilities

$$\begin{bmatrix} q & p & 0 & 0 & 0 & \cdot & \cdot & \cdot \\ q & 0 & p & 0 & 0 & \cdot & \cdot & \cdot \\ 0 & q & 0 & p & 0 & \cdot & \cdot & \cdot \\ 0 & 0 & q & 0 & p & \cdot & \cdot & \cdot \\ \cdot & \cdot & \cdot & \cdot & \cdot & \cdot & \cdot & \cdot \\ \cdot & \cdot & \cdot & \cdot & \cdot & \cdot & \cdot & \cdot \\ \cdot & \cdot & \cdot & \cdot & \cdot & \cdot & \cdot & \cdot \end{bmatrix}.$$

Notice the boundary condition at 0, where $r_0 = q$, $p_0 = p$. Here $\gamma_x = (q/p)^x$, and so the chain is transient if and only if $p > q$. If this is the case then $\rho_{xa} = (q/p)^{x-a}$ for $x > a$, and $\rho_{aa} = 2q$. Furthermore $m_{xb} = \dfrac{b - x}{p - q} - \dfrac{q}{(p - q)^2}[(q/p)^x - (q/p)^b]$ for $x < b$. On the other hand, if $q > p$ then the chain is recurrent and $m_{xa} = \dfrac{x - a}{q - p}$ for $x > a$, and $m_{aa} = \dfrac{q}{q - p}(q/p)^a$. If $p = q = \dfrac{1}{2}$ then the chain is recurrent and $m_{xb} = b(b + 1) - x(x + 1)$ for $x < b$, and $m_{xa} = \infty$ for $x \geq a$.

Let us next impose a right boundary condition on this chain, thereby confining it to $\{0, 1, \ldots, d\}$. Take the $(d + 1) \times (d + 1)$ transition matrix to be

$$\begin{bmatrix} q & p & 0 & 0 & 0 & \cdot & \cdot & \cdot \\ q & 0 & p & 0 & 0 & \cdot & \cdot & \cdot \\ 0 & q & 0 & p & 0 & \cdot & \cdot & \cdot \\ 0 & 0 & q & 0 & p & \cdot & \cdot & \cdot \\ \cdot & \cdot & \cdot & \cdot & \cdot & \cdot & \cdot & \cdot \\ \cdot & \cdot & \cdot & \cdot & \cdot & \cdot & \cdot & \cdot \\ \cdot & \cdot & \cdot & \cdot & \cdot & \cdot & \cdot & \cdot \\ \cdot & \cdot & \cdot & \cdot & \cdot & q & 0 & p \\ \cdot & \cdot & \cdot & \cdot & \cdot & 0 & q & p \end{bmatrix}.$$

This chain is always recurrent. Here if $p \neq q$ then for $x < b$, m_{xb} is given as earlier, and by symmetry for $x > a$, $m_{xa} = \dfrac{x - a}{q - p} - \dfrac{p}{(q - p)^2} \times$

$[(p/q)^{d-x} - (p/q)^{d-a}]$, and $m_{aa} = \dfrac{q}{p-q}(q/p)^a[(p/q)^{d+1} - 1]$. In the case $p = q$ we have for $x < b$, m_{xb} is given as earlier, and by symmetry for $x > a$, $m_{xa} = (d-a)(d-a+1) - (d-x)(d-x+1)$, and $m_{aa} = d + 1$.

Let us set up the gambler's ruin with an absorbing boundary at zero. The transition probabilities are given by

$$\begin{bmatrix} 1 & 0 & 0 & 0 & 0 & \cdot & \cdot & \cdot \\ q & 0 & p & 0 & 0 & \cdot & \cdot & \cdot \\ 0 & q & 0 & p & 0 & \cdot & \cdot & \cdot \\ 0 & 0 & q & 0 & p & \cdot & \cdot & \cdot \\ \cdot & \cdot & \cdot & \cdot & \cdot & \cdot & \cdot & \cdot \\ \cdot & \cdot & \cdot & \cdot & \cdot & \cdot & \cdot & \cdot \\ \cdot & \cdot & \cdot & \cdot & \cdot & \cdot & \cdot & \cdot \end{bmatrix}.$$

From the previous example, we infer that if $q > p$ then $\rho_{\{0\}}(x) = 1$ for all x, and $m_{x0} = \dfrac{x}{q-p}$ for $x > 0$. On the other hand if $q < p$ then $\rho_{\{0\}}(x) = (q/p)^x$. The two-player game would have the transition matrix

$$\begin{bmatrix} 1 & 0 & 0 & 0 & 0 & \cdot & \cdot & \cdot \\ q & 0 & p & 0 & 0 & \cdot & \cdot & \cdot \\ 0 & q & 0 & p & 0 & \cdot & \cdot & \cdot \\ 0 & 0 & q & 0 & p & \cdot & \cdot & \cdot \\ \cdot & \cdot & \cdot & \cdot & \cdot & \cdot & \cdot & \cdot \\ \cdot & \cdot & \cdot & \cdot & \cdot & \cdot & \cdot & \cdot \\ \cdot & \cdot & \cdot & \cdot & \cdot & \cdot & \cdot & \cdot \\ \cdot & \cdot & \cdot & \cdot & \cdot & q & 0 & p \\ \cdot & \cdot & \cdot & \cdot & \cdot & 0 & 0 & 1 \end{bmatrix}.$$

This time both 0 and d are absorbing states. We have $\rho_{\{0\}}(x) = \dfrac{(q/p)^d - (q/p)^x}{(q/p)^d - 1}$, $\rho_{xb} = \mathbb{P}_x(T_b < T_0) = \dfrac{(q/p)^x - 1}{(q/p)^b - 1}$ for $x < b$, $\rho_{xa} = \mathbb{P}_x(T_a < T_d) = \dfrac{(q/p)^d - (q/p)^x}{(q/p)^d - (q/p)^a}$ for $x > a$, and for $0 < a < d$ $\rho_{aa} = q\left[\dfrac{(q/p)^{a-1} - 1}{(q/p)^a - 1} + \dfrac{(q/p)^{d-a-1} - 1}{(q/p)^{d-a} - 1}\right]$. As regards the duration of the game, $\mathbf{E}_x(T_0 \wedge T_d) = \dfrac{x}{q-p} - \dfrac{d}{q-p}\dfrac{(q/p)^x - 1}{(q/p)^d - 1}$ for $0 < x < d$. In case $p = q$ these formulas are $\rho_{\{0\}}(x) = 1 - \dfrac{x}{d}$, $\rho_{xb} = \dfrac{x}{b}$ for $x < b$, $\rho_{xa} = \dfrac{d-x}{d-a}$ for $x > a$, $\rho_{aa} = 1 - \dfrac{1}{2}\dfrac{d}{a(d-a)}$ for $0 < a < d$ and $\mathbf{E}_x(T_0 \wedge T_d) = x(d-x)$ for $0 < x < d$.

For simple random walk on $\{0, \pm 1, \pm 2, \dots\}$ with no boundary, our previous analysis shows that if $p > q$ then $\rho_{xa} = (q/p)^{x-a}$ for $x > a$, $\rho_{aa} = 2q$ and $\mathbf{E}_x T_b = \dfrac{b-x}{p-q}$ for $x < b$.

Limiting Diffusions

We want to consider a birth and death chain that moves in times Δt, $2\Delta t$, $3\Delta t$, ... along the lattice state space $\{0, \pm\Delta x, \pm 2\Delta x, \dots\}$. Letting Δt, $\Delta x \to 0$ appropriately, we hope to generate a continuous movement $\{X(t)\colon t \geq 0\}$ on $\mathbb{R}$, called a *diffusion*. The transition probabilities $p_{k\Delta x}$, $r_{k\Delta x}$, $q_{k\Delta x}$ are allowed to depend on Δx; in order that the particle not run off to infinity in finite time or have discontinuous limiting trajectories, we require the balance that $p_{k\Delta x}(\Delta x) - q_{k\Delta x}(\Delta x) \to 0$ as $\Delta x \to 0$. Precisely, let us require that as $\Delta x \to 0$ and $k \to \infty$ in such a way that $k\Delta x \to x$, then

$$p_{k\Delta x}(\Delta x) - q_{k\Delta x}(\Delta x) = v(x)\Delta x + o(\Delta x),$$

and both $p_{k\Delta x}(\Delta x)$, $q_{k\Delta x}(\Delta x) \to \frac{1}{2}\sigma^2(x)$.

Our operator L now takes the form

$$L_{\Delta x}u(k\Delta x) = q_{k\Delta x}(\Delta x)u((k-1)\Delta x) - [p_{k\Delta x}(\Delta x) + q_{k\Delta x}(\Delta x)]u(k\Delta x) + p_{k\Delta x}(\Delta x)u((k+1)\Delta x).$$

Then as $\Delta x \to 0$ and $k \to \infty$ in such a way that $k\Delta x \to x$, we have

$$\frac{1}{(\Delta x)^2} L_{\Delta x}u(k\Delta x) \to \frac{1}{2}\sigma^2(x)\frac{d^2u(x)}{dx^2} + v(x)\frac{du(x)}{dx}.$$

That is, if we denote by L the differential operator

$$Lu(x) = \frac{1}{2}\sigma^2(x)\frac{d^2u(x)}{dx^2} + v(x)\frac{du(x)}{dx}$$

then $\dfrac{1}{(\Delta x)^2} L_{\Delta x}u \to Lu$. We refer to the function v as the *drift coefficient*, and to the function σ as the *diffusion coefficient*.

Let us examine what implications this has on the computation of the various functionals we considered earlier. Consider first the passage probabilities $u(x) = \mathbb{P}_x(T_a < T_b)$, $a < x < b$. This function must satisfy the second-order differential equation

$$Lu = 0,$$

with boundary conditions

$$u(a) = 1, \qquad u(b) = 0.$$

To analyze the time scale Δt consider the mean times $m(x) = \mathbf{E}_x(T_a \wedge T_b)$,

$a < x < b$. Before passing to the limit, we have the discrete difference equation

$$L_{\Delta x} m(k\Delta x) = -\Delta x.$$

In order to arrive at a finite limit we need to let $\Delta t \to 0$ in such a way that $\frac{\Delta t}{(\Delta x)^2} \to 1$. Then the function $m(x)$ will satisfy the limiting differential equation

$$Lm = -1,$$

with boundary conditions

$$m(a) = m(b) = 0.$$

Let us next consider imposition of a boundary at $x = 0$, so as to confine our process to the positive half line. Thus we set $q_0(\Delta x) \equiv 0$. We do not have to require that $p_0(\Delta x) \to 0$ at this left boundary point $x = 0$. Of course if $p_0(\Delta x) \equiv 0$ then the left boundary would be absorbing. Consider evaluating $w(x) = \mathbb{E}_x T_b$, $0 < x < b$. Again, assuming $\frac{\Delta t}{(\Delta x)^2} \to 1$, this function will satisfy the differential equation

$$Lw = -1,$$

with right boundary condition

$$w(b) = 0.$$

To come up with the appropriate constraint at the left boundary, note that for the discrete function we have

$$p_0(\Delta x)[w(\Delta x) - w(0)] = -\Delta t.$$

Thus by passing to the limit as $\Delta x \to 0$, we deduce that

$$\frac{d}{dx} w(0) = -\lim_{\Delta x \to 0} \frac{\Delta x}{p_0(\Delta x)}.$$

If this limit is infinite, then the existence of a finite solution w depends on the behavior of $v(x)$ and $\sigma^2(x)$ near $x = 0$.

The simplest and most important example of a limiting diffusion is *Brownian motion*. This is the limiting form of a simple random walk with $p = q = \frac{1}{2}$. The drift and diffusion coefficients are given by $v(x) \equiv 0$ and $\sigma(x) \equiv 1$, respectively. Another important example of a limiting diffusion is the *Ornstein-Uhlenbeck process*, which arises as the limiting form of the *Ehrenfest model*. In this model we have N balls, each of mass Δx, distributed among two urns. At each epoch, one ball is randomly chosen and removed from its urn and placed into the other urn. The state of the system is the total mass X in urn 1. As the individual masses Δx tend to zero, along with the time scale Δt, we hope to arrive at a continuous process in the limit. Observe here that

$$q_{k\Delta x}(\Delta x) = \frac{k}{N}, \qquad p_{k\Delta x}(\Delta x) = 1 - \frac{k}{N}.$$

Introduce the deviation

$$Y = X - \frac{1}{2}N\Delta x.$$

The transition probabilities for the Ys are

$$q_{k\Delta x}(\Delta x) = \frac{1}{2} + \frac{k}{N}, \qquad p_{k\Delta x}(\Delta x) = \frac{1}{2} - \frac{k}{N}.$$

We now send $\Delta x \to 0$ and k, $N \to \infty$ in such a way that $k\Delta x \to x$ and $N(\Delta x)^2 \to \lambda$. Then $v(x) = -\frac{2}{\lambda}x$ and $\sigma(x) \equiv 1$. The nature of v reveals the restoring mechanism inherent in this model.

Branching Chains

This is our first population model. In this model we assume that each individual in any generation gives rise to Z individuals in the next generation, where Z is a nonnegative integer-valued random variable having discrete density p_Z. The state X_n of the system at generation n is the total number of individuals in that generation. (Each individual lives for only one generation.) Thus

$$X_{n+1} = Z_1 + \cdots + Z_{X_n}, \tag{19}$$

where $Z_1, Z_2, \ldots$ are i.i.d. with common density p_Z. The process $\{X_n\}$ is a Markov chain, and its transition probabilities are given by

$$P(x, y) = \begin{cases} 0, & x = 0, \\ \mathbb{P}(Z_1 + \cdots + Z_x = y), & x \geq 1. \end{cases}$$

The state 0 is an absorbing state, and the event $X_n = 0$ corresponds to the population becoming extinct by the nth generation. In fact, the quantity most of interest concerning extinction is the individual *extinction probability* $\rho = \rho_{10}$, which is the probability of the line of a single individual becoming extinct. Since the individuals of any generation act independently in producing offspring, for any $x > 0$ we have $\rho_{x0} = \rho^x$. Because of the compound nature of X_{n+1} in (19) as a random sum, it is most convenient to study branching processes through generating functions. Indeed we see from (19) that

$$\Phi_{X_{n+1}}(t) = \Phi_{X_n}(\Phi_Z(t)),$$

and thus inductively

$$\Phi_{X_n} = \Phi_{X_0}(\Phi_Z^{(n)}),$$

where $\Phi_Z^{(n)}$ denotes the n-fold composite $\underbrace{\Phi_Z \circ \Phi_Z \circ \cdots \circ \Phi_Z}_{n \text{ times}}$. In particular $\mathbf{E}_1 t^{X_n} = \Phi_Z^{(n)}(t)$, and $P^n(x, y)$ is the coefficient of t^y in $[\Phi_Z^{(n)}(t)]^x$.

Whenever Z has finite expectation, $\mathbf{E}Z = \mu$, there holds

$$\mathbf{E}(X_n | X_0) = X_0 \mu^n. \tag{20}$$

This is proved inductively by conditioning as follows, using the Markov property:

$$\begin{aligned}\mathbf{E}(X_{n+1} | X_0) &= \mathbf{E}[\mathbf{E}(X_{n+1} | X_0, \ldots, X_n) | X_0] \\ &= \mathbf{E}[\mathbf{E}(X_{n+1} | X_n) | X_0] = \mu \mathbf{E}(X_n | X_0).\end{aligned}$$

Similarly, whenever Z has a second moment, if $\mathrm{Var}\, Z = \sigma^2$ there holds

$$\mathrm{Var}(X_n | X_0) = \begin{cases} X_0 \sigma^2 \mu^{n-1} \left(\dfrac{1 - \mu^n}{1 - \mu} \right), & \mu \neq 1, \\ X_0 n \sigma^2, & \mu = 1. \end{cases}$$

This is proved inductively, again using the Markov property, by conditioning as follows:

$$\begin{aligned}\mathrm{Var}(X_{n+1} | X_0) &= \mathrm{Var}[\mathbf{E}(X_{n+1} | X_0, \ldots, X_n) | X_0] \\ &\quad + \mathbf{E}[\mathrm{Var}(X_{n+1} | X_0, \ldots, X_n) | X_0] \\ &= \mathrm{Var}[\mathbf{E}(X_{n+1} | X_n) | X_0] + \mathbf{E}[\mathrm{Var}(X_{n+1} | X_n) | X_0] \\ &= \mu^2 \, \mathrm{Var}(X_n | X_0) + \sigma^2 \mathbf{E}(X_n | X_0) \\ &= \mu^2 \, \mathrm{Var}(X_n | X_0) + \sigma^2 X_0 \mu^n, \text{ by (20)}.\end{aligned}$$

The relevance of the generating function becomes even more profound when one observes that the extinction probability ρ satisfies

$$\begin{aligned}\rho = \rho_{10} &= P(1, 0) + \sum_{y=1}^{\infty} P(1, y) \rho_{y0} \\ &= P(1, 0) + \sum_{y=1}^{\infty} P(1, y) \rho^y = p_Z(0) + \sum_{y=1}^{\infty} p_Z(y) \rho^y = \Phi_Z(\rho).\end{aligned}$$

That is, ρ is a fixed point of Φ_Z. Before pursuing this, though, let us rule out the trivial case where $\Phi_Z(t) \equiv t$. This corresponds to Z being the unit mass at 1, so that $p_Z(1) = 1$. When this happens, X_n is constant in time, and the branching chain is not of interest. So let us assume now that

$$p_Z(1) < 1. \tag{21}$$

From this condition alone, it follows that $\mathscr{L}_R = \{0\}$ and $\mathscr{L}_T = \{1, 2, \ldots\}$;

since any state in $\mathscr{L}_T$ is visited only finitely often we arrive at the important dichotomy that either $X_n = 0$ in finite time or $X_n \to \infty$. The probability of the former event is of course our extinction probability ρ, when we start with a single individual ($X_0 = 1$).

Theorem VII. *The extinction probability ρ is the smallest positive fixed point of Φ_Z. If $\mu \leq 1$ then $\rho = 1$ and if $\mu > 1$ (including the possibility of infinity) then Φ_Z has a unique fixed point in $[0, 1)$.*

PROOF. The function Φ_Z is a nonnegative increasing function over the interval $[0, 1]$, and satisfies $\Phi_Z(1) = 1$. Furthermore Φ_Z' is increasing over $[0, 1)$ and satisfies $\mu = \lim_{t \uparrow 1} \Phi_Z'(t)$. Thus if $\mu < 1$ then $\Phi_Z' < 1$ on $[0, 1)$. Moreover even if $\mu = 1$, on account of (21) it is still true that $\Phi_Z' < 1$ on $[0, 1)$. Thus whenever $\mu \leq 1$

$$\frac{d}{dt}[\Phi_Z(t) - t] < 0, \qquad 0 \leq t < 1.$$

Since $\Phi_Z(1) = 1$ we conclude that

$$\Phi_Z(t) > t, \qquad 0 \leq t < 1.$$

Consider next what happens when $\mu > 1$. In this case there exists $t_0 \in (0, 1)$ such that $\Phi_Z' > 1$ on $(t_0, 1)$. By the Mean Value theorem

$$\frac{\Phi_Z(1) - \Phi_Z(t_0)}{1 - t_0} > 1,$$

and since $\Phi_Z(1) = 1$ we must have $\Phi_Z(t_0) < t_0$. Now since $\Phi_Z(0) \geq 0$ and Φ_Z is continuous there must be a fixed point ρ_0 of Φ_Z in $[0, t_0)$. Since $\mu > 1$ it follows that Φ_Z is strictly convex on $(0, 1]$, and thus Φ_Z can have only one fixed point in $[0, 1)$.

It remains only to show that when $\mu > 1$ the extinction probability ρ is equal to ρ_0, not 1 (the other fixed point). We use the facts that $\rho = \lim_n P^n(1, 0)$, and that $P^n(1, 0) = \Phi_Z^{(n)}(0)$. Since Φ_Z is increasing it follows inductively that $\Phi_Z^{(n)}(0) \leq \Phi_Z(\rho_0) = \rho_0$, and thus $\rho \leq \rho_0$. □

Queueing Chains

To describe various types of queues we use a notation proposed by Kendall. It is assumed that the successive interarrival times $T_1, T_2, \ldots$ between customers are independent identically distributed random variables. Similarly the service times $S_1, S_2, \ldots$ are assumed to be independent identically distributed random variables. One then writes a symbol of the form $F_T/F_S/Q$, where F_T denotes the d.f. of interarrival times, F_S denotes the d.f. of service times, and Q designates the number of servers.

The following symbols are used to denote interarrival and service time distributions:

D for a deterministic or constant interarrival or service time;
M for exponentially distributed interarrival or service time;
E_k for gamma (Erlangian) distributed interarrival or service time with parameters (k, λ);
G for a general distribution of service times;
GI for a general distribution of interarrival times.

Thus, for example, $M/G/1$ denotes a queue with exponentially distributed interarrival times, no special assumption about service times and one server; while $GI/M/1$ denotes a queue with no special assumptions about interarrival times, exponentially distributed service times, and one server. At this point in our studies we are restricted to considering queues that can be modeled so as to evolve in *discrete* time.

We begin our discussion with a model for a $GI/D/1$ queue. Suppose that if there are any customers waiting for service at the beginning of any given period, exactly one customer will be served during that period and that if there are no customers waiting for service at the beginning of a period, then none will be served during that period. Let $Z_1, Z_2, \ldots$ be an i.i.d. sequence of independent nonnegative integer-valued random variables having common discrete density p_Z. The Z_ns represent the number of customers arriving during the nth period. Let X_0 be the number of customers present initially, and let X_n denote the number of customers present at the end of period n. Thus if $X_n = 0$ then $X_{n+1} = Z_{n+1}$, and if $X_n \geq 1$ then $X_{n+1} = X_n + Z_{n+1} - 1$. The process $\{X_n\}$ is a Markov chain, and its transition probabilities are given by

$$P(x, y) = \begin{cases} p_Z(y), & x = 0, \\ p_Z(y - x + 1), & x \geq 1. \end{cases}$$

One of the important questions about a queue is whether or not it is recurrent.

Theorem VIII. *For the* GI/D/1 *chain,* $\rho = \rho_{00} = \rho_{10}$ *is the smallest nonnegative fixed point of* Φ. *Thus, an irreducible queueing chain is recurrent if and only if* $\mu = \mathbf{E}Z \leq 1$.

PROOF. We first prove that, as with the branching chain,

$$\rho_{x0} = \rho^x, \qquad x \geq 1. \tag{22}$$

Observe that if the chain starts at a positive integer x then $T_{x-1} = n$ if and only if

$$\begin{aligned} n &= \min(m > 0: x + (Z_1 - 1) + \cdots + (Z_m - 1) = x - 1) \\ &= \min(m > 0: Z_1 + \cdots + Z_m = m - 1). \end{aligned}$$

Thus $\mathbb{P}_x(T_{x-1} = n)$ is independent of x. In particular then $\rho_{x\,x-1}$ is independent of x, and $\rho_{x\,x-1} = \rho_{10} = \rho$. Since the chain is "left continuous,"

$$\rho_{x0} = \rho_{x\,x-1}\rho_{x-1\,x-2} \cdots \rho_{10},$$

thereby establishing (22). Relying on our analysis from the branching chains it follows that ρ is a fixed point of Φ_Z, and it only remains to show that if $\mu > 1$ then $\rho = \rho_0$ is the fixed point of Φ_Z which is less than one.

Observe that for any $n > 0$

$$\begin{aligned}\mathbb{P}_x(T_0 \le n) &\le \mathbb{P}_x(T_{x-1} \le n)\mathbb{P}_{x-1}(T_{x-2} \le n) \dots \mathbb{P}_1(T_0 \le n),\\ &= [\mathbb{P}_1(T_0 \le n)]^x.\end{aligned}$$

Since

$$\mathbb{P}_1(T_0 \le n+1) = p_Z(0) + \sum_{x=1}^{\infty} p_Z(x)\mathbb{P}_x(T_0 \le n)$$

we thus have

$$\mathbb{P}_1(T_0 \le n+1) \le \Phi_Z(\mathbb{P}_1(T_0 \le n)).$$

It follows now by induction that since Φ_Z is increasing on $[0, 1]$,

$$\mathbb{P}_1(T_0 \le n) \le \Phi_Z(\rho_0) = \rho_0.$$

Taking limits as $n \to \infty$ we conclude that $\rho \le \rho_0$. □

Another way to obtain discrete time queueing chains is by embedding continuous time queues. We describe first the embedded chain of an $M/G/1$ queue. Let X_n denote the number of persons waiting in line for service at the moment when the nth person to be served has finished service. That is, if $\{Y(t): t \ge 0\}$ is the continuous time $M/G/1$ process, and if $\{t_n\}$ are the moments right after successive customers depart, then we embed $X_n = Y(t_n)$. Let Z_n be the number of customers arriving during the time when the nth customer is being served. Then the discrete density for Z_n is given by

$$p_Z(x) = \int_0^{\infty} e^{-\lambda t}\frac{(\lambda t)^x}{x!}\, dF_S(t),$$

where $1/\lambda$ is the mean interarrival time. The connection between X_{n+1} and Z_{n+1} is exactly as earlier; namely, $X_{n+1} = Z_{n+1}$ if $X_n = 0$, and $X_{n+1} = X_n + Z_{n+1} - 1$ if $X_n > 0$.

We describe next the embedded $GI/M/1$ queue. Here we let X_n denote the number of persons waiting in line for service at the moment right before the nth customer arrives. Let Z_n denote the number of customers served during the interval between the arrivals of the nth and $(n+1)$st customers. The density of Z_n is given by

$$p_Z(x) = \int_0^\infty e^{-\lambda t}\frac{(\lambda t)^x}{x!}\,dF_T(t),$$

where $1/\lambda$ is now the mean service time. The connection between X_{n+1} and Z_n is that $X_{n+1} = \max(X_n + 1 - Z_n, 0)$. (The initial condition is $X_1 = 0$.) The process $\{X_n\}$ is again Markov, and its transition probabilities are given by

$$P(x, y) = \begin{cases} \mathbb{P}(Z \geq x + 1), & y = 0, \\ p_Z(x - y + 1), & y \geq 1. \end{cases}$$

Exercises

1. (Hoel, Port and Stone [28], Parzen [45]) For each of the following, find

(i) the decomposition of the state space;
(ii) the probabilities ρ_{xy};
(iii) the mean first passage times m_{xy};
(iv) $E_x(T_{\mathscr{L}_R})$, for each $x \in \mathscr{L}_T$.

(a) $\begin{bmatrix} 1 & 0 & 0 & 0 & 0 \\ \frac{1}{3} & 0 & \frac{2}{3} & 0 & 0 \\ 0 & \frac{1}{3} & 0 & \frac{2}{3} & 0 \\ 0 & 0 & \frac{1}{3} & 0 & \frac{2}{3} \\ 0 & 0 & 0 & 0 & 1 \end{bmatrix}$ (b) $\begin{bmatrix} 1 & 0 & 0 & 0 & 0 \\ \frac{1}{6} & \frac{1}{2} & \frac{1}{3} & 0 & 0 \\ 0 & \frac{1}{6} & \frac{1}{2} & \frac{1}{3} & 0 \\ 0 & 0 & \frac{1}{6} & \frac{1}{2} & \frac{1}{3} \\ 0 & 0 & 0 & 0 & 1 \end{bmatrix}$ (c) $\begin{bmatrix} 1 & 0 & 0 & 0 & 0 \\ \frac{1}{3} & 0 & \frac{2}{3} & 0 & 0 \\ 0 & \frac{1}{3} & 0 & \frac{2}{3} & 0 \\ 0 & 0 & \frac{1}{3} & 0 & \frac{2}{3} \\ 0 & 0 & 0 & \frac{1}{3} & \frac{2}{3} \end{bmatrix}$

(d) $\begin{bmatrix} \frac{1}{3} & \frac{2}{3} & 0 & 0 & 0 \\ \frac{1}{3} & 0 & \frac{2}{3} & 0 & 0 \\ 0 & \frac{1}{3} & 0 & \frac{2}{3} & 0 \\ 0 & 0 & \frac{1}{3} & 0 & \frac{2}{3} \\ 0 & 0 & 0 & 0 & 1 \end{bmatrix}$ (e) $\begin{bmatrix} 1 & 0 & 0 & 0 & 0 \\ \frac{1}{2} & 0 & \frac{1}{2} & 0 & 0 \\ 0 & \frac{1}{4} & \frac{1}{2} & \frac{1}{4} & 0 \\ 0 & 0 & \frac{1}{8} & \frac{3}{4} & \frac{1}{8} \\ 0 & 0 & 0 & 0 & 1 \end{bmatrix}$ (f) $\begin{bmatrix} \frac{1}{2} & \frac{1}{2} & 0 & 0 & 0 \\ \frac{1}{2} & 0 & \frac{1}{2} & 0 & 0 \\ 0 & \frac{1}{4} & \frac{1}{2} & \frac{1}{4} & 0 \\ 0 & 0 & \frac{1}{8} & \frac{3}{4} & \frac{1}{8} \\ 0 & 0 & 0 & 0 & 1 \end{bmatrix}$

(g) $\begin{bmatrix} \frac{1}{3} & \frac{2}{3} & 0 & 0 & 0 \\ \frac{1}{3} & 0 & \frac{2}{3} & 0 & 0 \\ 0 & \frac{1}{3} & 0 & \frac{2}{3} & 0 \\ 0 & 0 & \frac{1}{3} & 0 & \frac{2}{3} \\ 0 & 0 & 0 & \frac{1}{3} & \frac{2}{3} \end{bmatrix}$ (h) $\begin{bmatrix} 0 & \frac{2}{3} & 0 & 0 & \frac{1}{3} \\ \frac{1}{3} & 0 & \frac{2}{3} & 0 & 0 \\ 0 & \frac{1}{3} & 0 & \frac{2}{3} & 0 \\ 0 & 0 & \frac{1}{3} & 0 & \frac{2}{3} \\ \frac{2}{3} & 0 & 0 & \frac{1}{3} & 0 \end{bmatrix}$

(i) $\begin{bmatrix} 1 & 0 & 0 & 0 & 0 & 0 \\ 0 & 1 & 0 & 0 & 0 & 0 \\ \frac{1}{4} & 0 & \frac{1}{2} & 0 & 0 & \frac{1}{4} \\ 0 & \frac{1}{4} & 0 & \frac{1}{2} & 0 & \frac{1}{4} \\ 0 & 0 & 0 & 0 & 0 & 1 \\ \frac{1}{16} & \frac{1}{16} & \frac{1}{4} & \frac{1}{4} & \frac{1}{8} & \frac{1}{4} \end{bmatrix}$ (j) $\begin{bmatrix} \frac{1}{2} & \frac{1}{2} & 0 & 0 & 0 & 0 \\ \frac{1}{3} & \frac{2}{3} & 0 & 0 & 0 & 0 \\ 0 & 0 & \frac{1}{8} & 0 & \frac{7}{8} & 0 \\ \frac{1}{4} & \frac{1}{4} & 0 & 0 & \frac{1}{4} & \frac{1}{4} \\ 0 & 0 & \frac{3}{4} & 0 & \frac{1}{4} & 0 \\ 0 & \frac{1}{5} & 0 & \frac{1}{5} & \frac{1}{5} & \frac{2}{5} \end{bmatrix}$

$$\text{(k)} \begin{bmatrix} \frac{1}{3} & \frac{1}{3} & \frac{1}{3} & 0 & 0 & 0 & 0 \\ \frac{1}{3} & \frac{1}{3} & \frac{1}{3} & 0 & 0 & 0 & 0 \\ \frac{1}{3} & \frac{1}{3} & \frac{1}{3} & 0 & 0 & 0 & 0 \\ 0 & \frac{1}{4} & 0 & \frac{1}{2} & 0 & 0 & \frac{1}{4} \\ 0 & 0 & \frac{1}{4} & 0 & \frac{1}{2} & 0 & \frac{1}{4} \\ 0 & 0 & 0 & 0 & 0 & 0 & 1 \\ 0 & \frac{1}{16} & \frac{1}{16} & \frac{1}{4} & \frac{1}{4} & \frac{1}{8} & \frac{1}{4} \end{bmatrix} \qquad \text{(l)} \begin{bmatrix} \frac{1}{2} & 0 & \frac{1}{8} & \frac{1}{4} & \frac{1}{8} & 0 & 0 \\ 0 & 0 & 1 & 0 & 0 & 0 & 0 \\ 0 & 0 & 0 & 1 & 0 & 0 & 0 \\ 0 & 1 & 0 & 0 & 0 & 0 & 0 \\ 0 & 0 & 0 & 0 & \frac{1}{2} & 0 & \frac{1}{2} \\ 0 & 0 & 0 & 0 & \frac{1}{2} & \frac{1}{2} & 0 \\ 0 & 0 & 0 & 0 & 0 & \frac{1}{2} & \frac{1}{2} \end{bmatrix}$$

2. (Hoel, Port and Stone [28]) Consider a gene composed of d subunits, where d is some positive integer and each subunit is either normal or mutant in form. Consider a cell with a gene composed of m mutant subunits and $d - m$ subunits. Before the cell divides into two daughter cells, the gene duplicates. The corresponding gene of one of the daughter cells is composed of d units chosen at random from the $2m$ mutant subunits and the $2(d - m)$ normal subunits. Suppose we follow a fixed line of descent from a given gene. Let X_0 be the number of mutant subunits initially present, and let X_n, $n \geq 1$, be the number present in the nth descendant gene. Find $P(x, y)$ and compute $E(X_{n+1} | X_0, X_1, \ldots, X_n)$. Write the full matrix for the case $d = 5$.

3. (Hoel, Port and Stone [28]) Suppose we have two boxes and $2d$ balls, of which d are black and d are red. Initially, d of the balls are placed in box 1, and the remainder of the balls are placed in box 2. At each trial a ball is chosen at random from each of the boxes, and the two balls are put back in the opposite boxes. Let X_0 denote the number of black balls initially in box 1 and, for $n \geq 1$, let X_n denote the number of black balls in box 1 after the nth trial. Find $P(x, y)$. Write the full matrix for the case $d = 5$.

4. (Hoel, Port and Stone [28]) Consider a Markov chain on the nonnegative integers such that, starting from x, the chain goes to state $x + 1$ with probability p, $0 < p < 1$, and goes to state 0 with probability $1 - p$. Find the decomposition of the state space, and compute $P_0(T_0 = n)$.

5. (Hoel, Port and Stone [28]) Prove: If $y \in \mathscr{L}_T$ then

$$\sum_{n=0}^{\infty} P^n(x, y) \leq \sum_{n=0}^{\infty} P^n(y, y), \qquad x \in \mathscr{L}.$$

6. (Hoel, Port and Stone [28])

$$P(x, y) = \binom{d}{y} \left(\frac{x}{d}\right)^y \left(1 - \frac{x}{d}\right)^{d-y} \qquad 0 \leq x, y \leq d.$$

Find $\rho_{\{0\}}(x)$.

7. (Parzen [45]) A white rat is put into the maze below:

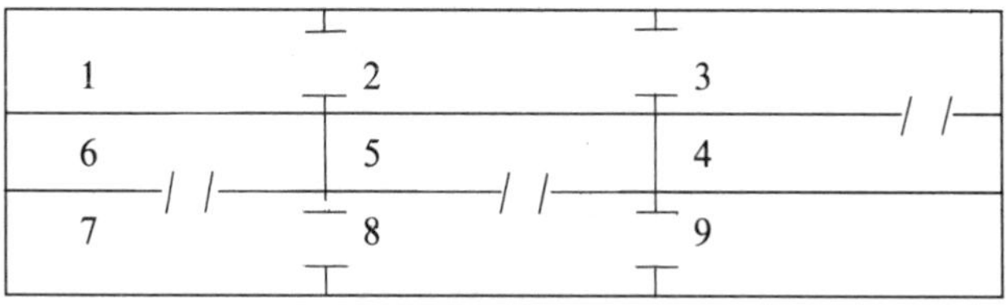

1 2 3
6 5 4
7 8 9

The rat moves through the compartments at random; i.e., if there are k ways to leave a compartment he chooses one with probability $\frac{1}{k}$. He makes one change of compartment at each instant of time. The state of the system is the number of the compartment the rat is in. Find $E_1 T_1$, $E_1 T_1^2$, $E_1 T_7$, $E_1 T_7^2$.

8. In the so-called brother-sister mating two individuals are mated, and among their direct descendants two individuals of opposite sex are selected at random. These are again mated, and the process continues indefinitely. With three genotypes AA, Aa, aa for each parent, we have to distinguish six combinations of parents which we label as follows: $E_1 = AA \times AA$, $E_2 = AA \times Aa$, $E_3 = Aa \times Aa$, $E_4 = Aa \times aa$, $E_5 = aa \times aa$, $E_6 = AA \times aa$. Set up the transition matrix.

9. Let $Z_1, Z_2, \ldots$ be the outcomes from the successive rolls of a die, and set

$$X_n = \max(Z_i: 1 \leq i \leq n).$$

Find the powers P^m.

10. Let $Z_1, Z_2, \ldots$ be a sequence of Bernoulli trials and set

$$X_n = \begin{cases} 1 & Z_{n-1}, Z_n = SS \\ 2 & Z_{n-1}, Z_n = SF \\ 3 & Z_{n-1}, Z_n = FS \\ 4 & Z_{n-1}, Z_n = FF \end{cases} \qquad (n \geq 2)$$

Set up the transition matrix.

11. (Hoel, Port and Stone [28]) A gambler playing roulette makes a series of one dollar bets. He has respective probabilities $\frac{9}{19}$ and $\frac{10}{19}$ of winning and losing each bet. The gambler decides to quit playing as soon as his winnings reach 25 dollars or his net losses reach 10 dollars.

(a) Find the probability that when he quits playing he will have won 25 dollars.
(b) Find his expected loss.

12. (Hoel, Port and Stone [28]) Let

$$p_x = \frac{x+2}{2(x+1)}, \qquad q_x = \frac{x}{2(x+1)} \qquad x = 0, 1, 2, \ldots.$$

Determine whether the chain is recurrent or transient. Compute

(a) $P_x(T_a < T_b)$, for $a < x < b$.
(b) ρ_{x0}, for $x > 0$.

13. (Hoel, Port and Stone [28]) Suppose $\frac{q_x}{p_x} = \left(\frac{x}{x+1}\right)^2$, $x \geq 1$. Determine whether the chain is recurrent or transient. Compute ρ_{x0}, for $x > 0$.

14. (Hoel, Port and Stone [28]) Suppose the gambler in Problem 11 quits playing as soon as he either is one dollar ahead or has lost his initial capital of \$1000.

(a) Find the probability that when he quits playing he will have lost \$1000.
(b) Find his expected loss.

15. (Hoel, Port and Stone [28]) Find the mean first passage times $E_x T_x$, $0 \le x \le d$, for the Ehrenfest chain.

16. (Hoel, Port and Stone [28]) Modified Ehrenfest chain: This time the (random) ball that is selected is returned to a random box. Find the mean first passage times $E_x T_x$, $0 \le x \le d$.

17. (Hoel, Port and Stone [28]) Find the mean first passage times $E_x T_x$, $0 \le x \le d$, for Problem 3 (the red and black balls with random switching).

18. (Parzen [45]) Consider a sequence of independent tosses of a coin that has probability p of falling heads. Let X_n be the length of the success run underway at the time of the nth trial. That is, for $n \ge 1$, $X_n = k$ if the outcome in the $(n - k)$th trial was tails and the outcomes of trials $n - k + 1$, $n - k + 2$, $\ldots$, n were all heads. Find ρ_{x0}, $x \ge 0$.

19. (Karlin and Taylor [31]) Simple Random walk:

    (a) Determine $P^n(0, 0)$.
    (b) Find the generating function

$$\sum_{n=0}^{\infty} P^n(0, 0)x^n.$$

    (c) Find the generating function

$$\mathbf{E}_0 x^{T_0}.$$

    (d) What is ρ_{00}?

20. Solve for $u(x)$, $m(x)$ on pages 89, 90.

21. (Hoel, Port and Stone [45]) Suppose that every man in a certain society has exactly three children, who independently have probability one-half of being a boy and one-half of being a girl. Find the probability that the male line of a given man eventually becomes extinct.

    If a given man has two boys and one girl, what is the probability that his male line will continue forever?

22. Suppose that every man in a certain society stops having children as soon as he has one girl, and independently every child has probability p of being a girl. Find the probability that the male line of a given man eventually becomes extinct. What if every man stops having children as soon as he has two girls?

23. (Hoel, Port and Stone [45]) Find necessary and sufficient conditions for the queueing chain on page 94 to be irreducible. Find $\mathscr{L}_R$ and $\mathscr{L}_T$ for those cases when it is reducible.

24. Consider a sequence of independent trials with two possible outcomes, S or F. Suppose that in trial n the probability of S is p_n. Let X_n be the length of the success run after trial n. That is, $X_n = k$ if the outcomes in the preceding trials were

Trial	$n-k$	$n-k+1$	$n-k+2 \ldots n$
Outcome	F	S	$S \quad \ldots S$

$$|\longleftarrow k \longrightarrow|$$

Find necessary and sufficient conditions for this chain to be recurrent.

25. (Karlin and Taylor [31]) The following sequential approach to estimating the size of a finite population, perhaps a wildlife population such as fish, is proposed. A member of the population is sampled at random, tagged, and returned. Another is sampled, tagged, and returned, and so on, until a member is drawn that has been before. When this occurs, say, at trial T, we stop (and possibly begin again with a new kind of tag). Suppose the population size is N.

    (a) Compute the pdf of T.
    (b) How would you estimate N from T?

26. (Parzen [45]) Consider the queueing chain on page 96. Find necessary and sufficient conditions that it be recurrent.

27. (Karlin and Taylor [31]) Suppose $\mathscr{L} = \{0, 1, \ldots, d\}$ and

$$P(x, y) = \binom{d}{y} \alpha_x^y (1 - \alpha_x)^{d-y}$$

where

$$\alpha_x = \frac{1 - \varepsilon^x}{1 - \varepsilon^d} \qquad (0 \le \varepsilon \le 1)$$

Observe that 0, d are absorbing states. Compute $\rho_{\{d\}}(x)$.

28. (Karlin and Taylor [31]) Consider a random walk on a circle on which four points have been marked. Let the probabilities of moving clockwise and counterclockwise be p and q, respectively. Find the mean first passage times.

29. Consider a random walk on a circle on which $d + 1$ points $(0, 1, \ldots, d)$ have been marked. Let the probabilities of moving clockwise and counterclockwise be p and q, respectively. Find necessary and sufficient conditions for this chain to be irreducible. Assuming it is irreducible, find the mean first passage times.

30. Consider two-dimensional symmetric random walk in the full infinite plane. (That is, the probabilities of a transition one unit north, east, south, west are all $\frac{1}{4}$.) Find $P^n(\mathbf{0}, \mathbf{0})$, where $\mathbf{0} = (0, 0)$ is the origin. Use Stirling's formula

$$n! \sim n^{n+(1/2)} e^{-n} \sqrt{2\pi}$$

to conclude that this chain is recurrent. What about three-dimensional random walk? (See Karlin and Taylor [31], pages 67–69.)

Section V
Markov Chains—Stationary Distributions and Steady State*

Stationary Distributions

Let $N_n(y)$ denote the number of visits of the Markov chain $\{X_n\}$ to y during times $m = 1, \ldots, n$. That is,

$$N_n(y) = \sum_{m=1}^{n} I_{\{y\}}(X_m).$$

Theorem I. *For* $y \in \mathscr{L}_R$

$$\lim_n \frac{N_n(y)}{n} = \frac{1}{m_{yy}} I_{\{T_y < \infty\}} \qquad \text{a.s.} \tag{1}$$

and

$$\lim_n \mathbf{E}_x \frac{N_n(y)}{n} = \frac{\rho_{xy}}{m_{yy}}. \tag{2}$$

PROOF. For $r \geq 1$ let $T_y^{(r)}$ denote the time of the rth visit to y.

$$T_y^{(r)} = \min(n \geq 1 : N_n(y) = r).$$

Set $W_y^{(1)} = T_y$ and $W_y^{(r)} = T_y^{(r)} - T_y^{(r-1)}$, $r \geq 2$. Then $W_y^{(r)}$ is the *waiting time* between the $(r-1)$st visit and rth visit to y. We claim that the random variables $W_y^{(2)}, W_y^{(3)}, \ldots$ are i.i.d. Indeed

$$\mathbb{P}(W_y^{(r+1)} = m_{r+1} \mid W_y^{(1)} = m_1, \ldots, W_y^{(r)} = m_r) = \mathbb{P}_y(T_y = m_{r+1}).$$

Now $T_y^{(r)} = W_y^{(1)} + \cdots + W_y^{(r)}$, and thus by the Strong Law of Large Numbers, a.s.

* Material taken from Hoel, Port and Stone [28] with permission.

$$\lim_{r\to\infty} \frac{T_y^{(r)}}{r} = \begin{cases} m_{yy}, & T_y < \infty, \\ \infty & T_y = \infty. \end{cases}$$

If $T_y < \infty$ then $N_n(y) \to \infty$, and since

$$T_y^{(N_n(y))} \le n < T_y^{(N_n(y)+1)}$$

we obtain (for $n \ge T_y$)

$$\frac{T_y^{(N_n(y))}}{N_n(y)} \le \frac{n}{N_n(y)} < \frac{T_y^{(N_n(y)+1)}}{N_n(y)}.$$

Consequently a.s.

$$\lim_n \frac{N_n(y)}{n} = \begin{cases} \dfrac{1}{m_{yy}}, & T_y < \infty, \\ 0, & T_y = \infty. \end{cases}$$

which establishes (1). To obtain (2) use the Dominated Convergence Theorem, since $0 \le \dfrac{N_n(y)}{n} \le 1$. □

There is an interesting alternate proof of (2) that relies on the following lemma due to Abel, stated here without proof.

Abel's Lemma. *Let* $\{a_n\}$ *be a nonnegative sequence of real numbers with generating function*

$$\Phi(t) = \sum_{n=0}^{\infty} a_n t^n, \qquad |t| < 1.$$

There exists a finite limit

$$\lim_n \frac{1}{n} \sum_{m=1}^{n} a_m = L$$

if and only if

$$\lim_{t\to 1-} (1-t)\Phi(t) = L.$$

ALTERNATE PROOF OF (2) (from Parzen [45]). Set

$$P(t; x, y) = \sum_{n=0}^{\infty} P^n(x, y) t^n.$$

By Abel's lemma it suffices to show that

$$\lim_{t\to 1-} (1-t) P(t; x, y) = \frac{\rho_{xy}}{m_{yy}}. \tag{3}$$

Observe that for $n \ge 1$

$$P^n(x, y) = \sum_{m=1}^{n} \mathbb{P}_x(T_y = m)P^{n-m}(y, y).$$

Thus if we let Φ be the generating function

$$\Phi(t; x, y) = \mathbf{E}_x t^{T_y} = \sum_{n=1}^{\infty} \mathbb{P}_x(T_y = n)t^n,$$

then

$$\Phi(t; x, y)P(t; y, y) = \begin{cases} P(t; y, y) - 1, & x = y, \\ P(t; x, y), & x \neq y. \end{cases}$$

We can use this to express P in terms of Φ as follows:

$$(1 - t)P(t; x, y) = \begin{cases} \dfrac{1 - t}{1 - \Phi(t; y, y)}, & x = y, \\ \dfrac{1 - t}{1 - \Phi(t; y, y)}\Phi(t; x, y), & x \neq y. \end{cases}$$

Since

$$\lim_{t \to 1-} \frac{1 - \Phi(t; y, y)}{1 - t} = \frac{d}{dt}\Phi(1; y, y) = m_{yy}$$

and

$$\lim_{t \to 1-} \Phi(t; x, y) = \rho_{xy},$$

we obtain (3). □

A recurrent state y is called *null recurrent* if $m_{yy} = \infty$. Otherwise, if $m_{yy} < \infty$ it is called *positive recurrent*. Denote the sets of null recurrent and positive recurrent states by $\mathscr{L}_N$ and $\mathscr{L}_P$, respectively.

Theorem II. *$\mathscr{L}_P$ and $\mathscr{L}_N$ are closed. If $\mathscr{L}$ is finite then $\mathscr{L}_N = \phi$.*

Proof. Let $x \in \mathscr{L}_P$ and suppose x leads to y. Then y leads to x and we can find k, l such that $P^k(x, y) > 0$ and $P^l(y, x) > 0$. Since for any $m \geq 0$

$$P^{m+k+l}(y, y) \geq P^l(y, x)P^m(x, x)P^k(x, y)$$

we have

$$\frac{1}{n}\sum_{m=k+l+1}^{k+l+n} P^m(y, y) \geq P^l(y, x)P^k(x, y)\frac{1}{n}\sum_{m=1}^{n} P^m(x, x).$$

Taking limits as $n \to \infty$ we deduce that

$$\frac{1}{m_{yy}} \geq P^l(y, x)P^k(x, y)\frac{1}{m_{xx}} > 0.$$

Thus $y \in \mathscr{L}_P$.

If C is a finite closed set then it must contain a positive recurrent state. Otherwise we obtain, for $x \in C$,

$$1 = \lim_n \sum_{y \in C} \frac{1}{n} \sum_{m=1}^{n} P^m(x, y) = \sum_{y \in C} \lim_n \frac{1}{n} \sum_{m=1}^{n} P^m(x, y) = 0. \qquad \square$$

We have already discovered an analytical way of classifying states. If $y \in \mathscr{L}_T$ then $\sum_{n=0}^{\infty} P^n(y, y) = \frac{1}{1 - \rho_{yy}} < \infty$, whereas if $y \in \mathscr{L}_R$ then $\sum_n P^n(y, y) = \infty$. If $y \in \mathscr{L}_P$ then $\lim_n \frac{1}{n} \sum_{m=1}^{n} P^m(y, y) = \frac{1}{m_{yy}} > 0$, whereas if $y \in \mathscr{L}_N$ then

$$\lim_n \frac{1}{n} \sum_{m=1}^{n} P^m(y, y) = 0.$$

Regarding the terms $P^n(x, y)$ the same limits apply, but multiplied by ρ_{xy}. Thus if $y \in \mathscr{L}_T$ then for any $x \in \mathscr{L}$ $\sum_{n=0}^{\infty} P^n(x, y) = \frac{\rho_{xy}}{1 - \rho_{yy}}$, whereas if $y \in \mathscr{L}_R$ then $\sum_n P^n(x, y) = \infty$ or 0, depending on whether or not $\rho_{xy} > 0$, respectively. If $y \in \mathscr{L}_P$ then for any $x \in \mathscr{L}$, $\lim_n \frac{1}{n} \sum_{m=1}^{n} P^m(x, y) = \frac{\rho_{xy}}{m_{yy}}$, whereas if $y \in \mathscr{L}_N$ then $\lim_n \frac{1}{n} \sum_{m=1}^{n} P^m(x, y) = 0$. One thing we note is that the entire array $\frac{1}{n} \sum_{m=1}^{n} P^m$ has a limit. The analysis of the limiting behavior of P^n is more delicate. Of course if $y \in \mathscr{L}_T$ then for any x we have $\lim_n P^n(x, y) = 0$, since $P^n(x, y)$ is summable over n. It can be shown that if $y \in \mathscr{L}_N$ then again $\lim_n P^n(x, y) = 0$ for any x, although $P^n(x, y)$ is not summable over n (whenever $\rho_{xy} > 0$). The analysis of the asymptotics of $P^n(x, y)$ for large n when $y \in \mathscr{L}_P$ is developed later. Of course if $\lim_n P^n(x, y)$ exists then it must be equal to $\frac{\rho_{xy}}{m_{yy}}$, but as we shall see later this limit need not exist. Nevertheless, we shall still discover the behavior of $P^n(x, y)$ for large n.

A *stationary distribution* π is a discrete density $\{\pi(x) : x \in \mathscr{L}\}$ satisfying

$$\pi P = \pi.$$

Here, as earlier, we treat π as a (finite or countably infinite) row vector. We showed in the beginning of Section IV that if π_n denotes the density of X_n then $\pi_n = \pi_0 P^n$. Thus π_n is independent of n if and only if π_0 is stationary.

Theorem III. *An irreducible positive recurrent Markov chain has a unique stationary distribution* π, *given by*

$$\pi(x) = \frac{1}{m_{xx}}, \qquad x \in \mathscr{L}.$$

PROOF. Suppose first that π is any stationary distribution. Then for any $x \in \mathscr{L}$

$$\sum_{z \in \mathscr{L}} \pi(z) \left[\frac{1}{n} \sum_{m=1}^{n} P^m(z, x) \right] = \pi(x).$$

Taking limits as $n \to \infty$ shows that

$$\pi(x) = \frac{1}{m_{xx}}, \qquad x \in \mathscr{L}.$$

This establishes uniqueness. It also shows that if a stationary distribution exists, then $\sum_x \frac{1}{m_{xx}} = 1$.

Observe next that for any finite subset $\mathscr{L}_1 \subseteq \mathscr{L}$ and any $z \in \mathscr{L}$

$$\sum_{x \in \mathscr{L}_1} \left[\frac{1}{n} \sum_{m=1}^{n} P^m(z, x) \right] \leq 1,$$

and thus letting $n \to \infty$

$$\sum_{x \in \mathscr{L}_1} \frac{1}{m_{xx}} \leq 1.$$

Since $\mathscr{L}_1$ is arbitrary we conclude that

$$\sum_{x \in \mathscr{L}} \frac{1}{m_{xx}} \leq 1.$$

Similarly

$$\sum_{x \in \mathscr{L}_1} \left[\frac{1}{n} \sum_{m=1}^{n} P^m(z, x) \right] P(x, y) \leq \frac{1}{n} \sum_{m=1}^{n} P^{m+1}(z, y)$$

and thus

$$\sum_{x \in \mathscr{L}} \frac{1}{m_{xx}} P(x, y) \leq \frac{1}{m_{yy}}, \qquad y \in \mathscr{L}. \tag{4}$$

Summing over y we find that

$$\sum_{y \in \mathscr{L}} \frac{1}{m_{yy}} \geq \sum_{y \in \mathscr{L}} \sum_{x \in \mathscr{L}} \frac{1}{m_{xx}} P(x, y) = \sum_{x \in \mathscr{L}} \frac{1}{m_{xx}},$$

and thus equality must attain in (4) for every $y \in \mathscr{L}$. This establishes existence. □

Theorem IV. *Let π be a stationary distribution for a Markov chain. If* $x \notin \mathscr{L}_P$ *then* $\pi(x) = 0$. *In particular, an irreducible Markov chain is positive recurrent if and only if it has a stationary distribution.*

PROOF. Observe that

$$\sum_{x \in \mathscr{L}} \pi(z)\left[\frac{1}{n}\sum_{m=1}^{n} P^m(z, x)\right] = \pi(x).$$

If $x \notin \mathscr{L}_P$ then

$$\lim_n \frac{1}{n}\sum_{m=1}^{n} P^m(z, x) = 0.$$

and thus by the Dominated Convergence Theorem $\pi(x) = 0$. □

Observe now that for an irreducible positive recurrent Markov chain with stationary distribution π

$$\lim_n \frac{N_n(x)}{n} = \pi(x) \qquad \text{a.s.,} \qquad x \in \mathscr{L}.$$

Theorem V. (i) *If* $\mathscr{L}_P = \phi$ *then no stationary distribution exists.*

(ii) *If* $\mathscr{L}_P \neq \phi$ *is irreducible then there is a unique stationary distribution.*

(iii) *If* $\mathscr{L}_P \neq \phi$ *is reducible then there are infinitely many stationary distributions.*

PROOF. If C is a closed irreducible positive recurrent state then it has a stationary distribution $\pi_C(x)$, $x \in C$. If we extend π_C to all of $\mathscr{L}$ as

$$\hat{\pi}_C(x) = \begin{cases} \pi_C(x), & x \in C, \\ 0 & x \notin C, \end{cases}$$

then $\hat{\pi}_C$ is a stationary distribution on $\mathscr{L}$. Furthermore any stationary distribution is supported on $\mathscr{L}_p$. Thus if $\mathscr{L}_P$ is irreducible, $\mathscr{L}_P = C$, then $\hat{\pi}_C$ is the unique stationary distribution. Otherwise if $\mathscr{L}_P = \bigcup_k C_k$ where the C_ks are closed irreducible then any convex combination of the $\hat{\pi}_{C_k}$s is a stationary distribution, and distinct convex combinations correspond to distinct stationary distributions. □

Observe from this proof that if C is any closed irreducible positive recurrent set, then there is a unique stationary distribution on $\mathscr{L}$ that is supported on C.

Let $x \in \mathscr{L}$ be such that $\rho_{xx} > 0$ and set

$$d_x = \text{g.c.d.}(n \geq 1 : P^n(x, x) > 0).$$

Theorem VI. *If* x *and* y *lead to one another then* $d_x = d_y$.

PROOF. Choose k and l such that

$$P^k(x, y) > 0, \qquad P^l(y, x) > 0.$$

Then $d_x | k + l$. Furthermore for any $n \geq 1$

$$P^{k+l+n}(x, x) \geq P^k(x, y)P^n(y, y)P^l(y, x),$$

and so

$$P^n(y, y) > 0 \Rightarrow d_x | k + l + n \Rightarrow d_x | n.$$

This implies that $d_x | d_y$. □

If the Markov chain is irreducible, then all states have a common d, referred to as the *period* of the chain. If $d = 1$ the chain is said to be *aperiodic.*

Theorem VII. *If* $\{X_n : n \geq 0\}$ *is an aperiodic irreducible Markov chain then for every pair* $x, y \in \mathscr{L}$ *there exists* $n_0 = n_0(x, y)$ *such that* $P^n(x, y) > 0$ *for all* $n \geq n_0$.

PROOF. If suffices to treat the case $x = y$, since the chain is irreducible. Let

$$I = \{n > 0 : P^n(x, x) > 0\}.$$

Then (i) g.c.d.$(I) = 1$, and (ii) I is closed under addition.

We first establish that I contains two consecutive integers. If not there would have to be a *minimal* difference $k \geq 2$ between any two integers of I. Let n_1, $n_1 + k \in I$. Since g.c.d.$(I) = 1$ there exists $n \in I$ that is not a multiple of k, say $n = mk + r$, $0 < r < k$. Since I is closed under addition $(m + 1)(n_1 + k)$ and $n + (m + 1)n_1$ belong to I. But their difference is then

$$(m + 1)(n_1 + k) - n - (m + 1)n_1 = k - r < k,$$

which is a contradiction. Therefore $k = 1$, as claimed.

Let n_1, $n_1 + 1$ be two consecutive integers in I. Let $n \geq n_0 = n_1^2$, and write $n - n_1^2 = mn_1 + r$, $0 \leq r < n_1$. Then

$$n = r(n_1 + 1) + (n_1 - r + m)n_1 \in I.$$ □

Theorem VIII. *Let* $\{X_n : n \geq 0\}$ *be an irreducible positive recurrent Markov chain having stationary distribution* π *and period* d. *For each pair* $x, y \in \mathscr{L}$ *there corresponds* $r = r(x, y)$, $0 \leq r < d$ *such that*

$$\lim_m P^{md+k}(x, y) = \begin{cases} d\pi(y), & k = r, \\ 0, & 0 \leq k < d, k \neq r. \end{cases}$$

In particular if the chain is aperiodic then for any $x, y \in \mathscr{L}$

$$\lim_n P^n(x, y) = \pi(y). \tag{5}$$

PROOF. We first treat the case $d = 1$ when the chain is aperiodic. Define a Markov chain $\{(X_n, Y_n): n \geq 0\}$ on $\mathscr{L} \times \mathscr{L}$ so that the marginals $\{X_n\}$ and $\{Y_n\}$ are each Markov chains having transition probabilities P, and making independent transitions. That is,

$$P_2((x, y), (x', y')) = P(x, x')P(y, y').$$

Since

$$P_2^n((x, y), (x', y')) = P^n(x, x')P^n(y, y'),$$

it follows from the previous theorem that $P_2^n((x, y), (x', y')) > 0$ for n sufficiently large. Therefore $\{(X_n, Y_n)\}$ is irreducible. Furthermore $\pi(x, y) = \pi(x)\pi(y)$ is a stationary distribution for $\{(X_n, Y_n)\}$, and thus this chain is also recurrent.

Let $C = \{(x, x): x \in \mathscr{L}\}$. Observe that for $n \geq m$

$$\begin{aligned} \mathbb{P}(X_n = y \mid T_C = m, X_m = Y_m = z) &= P^{n-m}(z, y) \\ &= \mathbb{P}(Y_n = y \mid T_C = m, X_m = Y_m = z). \end{aligned}$$

Thus summing over m and z

$$\mathbb{P}(X_n = y \mid T_C \leq n) = \mathbb{P}(Y_n = y \mid T_C \leq n).$$

From this, it follows that

$$\begin{aligned} \mathbb{P}(X_n = y) &= \mathbb{P}(Y_n = y, T_C \leq n) + \mathbb{P}(X_n = y, T_C > n) \\ &\leq \mathbb{P}(Y_n = y) + \mathbb{P}(T_C > n) \end{aligned}$$

Reversing the roles of X_n and Y_n leads to the estimate

$$|\mathbb{P}(X_n = y) - \mathbb{P}(Y_n = y)| \leq \mathbb{P}(T_C > n) \tag{6}$$

Since $\{(X_n, Y_n)\}$ is irreducible recurrent $T_C < \infty$ a.s., and so the right-hand side of (6) tends to zero as $n \to \infty$. By initializing X_0 according to the unit mass δ_x, and initializing Y_0 according to π we obtain (5). This concludes the proof for the aperiodic case.

For the case when $d > 1$ set

$$\hat{X}_n = X_{nd}, \qquad n \geq 0.$$

Then $\{\hat{X}_n : n \geq 0\}$ is a Markov chain with transition probabilities $\hat{P} = P^d$, and all states have period one with respect to the $\hat{X}_n$ chain. Since $\hat{m}_{yy} = \frac{1}{d} m_{yy}$ we conclude that

$$\lim_n \hat{P}(y, y) = \frac{d}{m_{yy}} = d\pi(y). \tag{7}$$

Next let x, y be any pair of states and choose l such that $P^l(y, x) > 0$. For any $n \geq 0$

$$P^{n+l}(x, x) \geq P^n(x, y)P^l(y, x),$$

and thus

$$P^n(x, y) > 0 \Rightarrow n = -l \qquad (\text{mod } d).$$

Choose $0 \leq r < d$ so that $r = -l$ (mod d). Since

$$P^{nd+r}(x, y) = \sum_{k=0}^{n} \mathbb{P}_x(T_y = kd + r)P^{(n-k)d}(y, y)$$

we obtain from (7) that

$$\lim_n P^{nd+r}(x, y) = d\pi(y). \qquad \square$$

We want to analyze $P^n(x, y)$ for large n, when $x \in \mathscr{L}_T$ and $y \in \mathscr{L}_P$. It follows from our preceding result that if y and z are in the same closed irreducible part C of $\mathscr{L}_P$ then there is a unique $r = r(z, y)$, $0 \leq r < d$, such that $P^n(z, y) > 0 \Rightarrow n = r$ (mod d), where $d = d_y$ is the (common) period in C. Based on getting to y, then, we can partition C into pieces $C(0), \ldots, C(d-1)$ such that $z \in C(k)$ if and only if $r(z, y) = k$. When $x \in \mathscr{L}_T$ and $y \in C$ then even if $d_y > 1$ it may happen that $P^n(x, y) > 0$ for all n. More of interest in this case is $\mathbb{P}_x(T_y = r \text{ (mod } d))$. If we set up vectors

$$u_r(x) = \mathbb{P}_x(T_y = r \text{ (mod } d)), \qquad x \in \mathscr{L}_T, 0 \leq r < d,$$

then we can compute these probabilities implicitly through the systems of linear equations

$$u_{r+1} = P_T u_r + f_r, \qquad 0 \leq r < d,$$

where f_r is the vector

$$f_r(x) = \mathbb{P}_x(C(r)), \qquad x \in \mathscr{L}_T, 0 \leq r < d,$$

P_T is P restricted to $\mathscr{L}_T$ and $u_d = u_0$.

Theorem IX. *Let* $\mathrm{x} \in \mathscr{L}_\mathrm{T}$ *and* $\mathrm{y} \in \mathrm{C}$, *where* C *is a closed irreducible subset of* $\mathscr{L}_\mathrm{P}$, *and let* $\mathrm{d} = \mathrm{d_y}$ *be the period of the states in* C. *Then for* $0 \leq \mathrm{r} < \mathrm{d}$

$$\lim_n P^{nd+r}(x, y) = d\pi_C(y)\mathbb{P}_x(T_y = r \text{ (mod } d)).$$

PROOF.

$$\begin{aligned}P^{nd+r}(x, y) &= \sum_{k=0}^{n} \mathbb{P}_x(T_y = kd + r)P^{(n-k)d}(y, y)\\ &= d\pi_C(y)\mathbb{P}_x(T_y = r \text{ (mod } d), T_y \leq nd + r)\\ &\quad + \sum_{k=0}^{n} \mathbb{P}_x(T_y = kd + r)[P^{(n-k)d}(y, y) - d\pi_C(y)]. \qquad \square\end{aligned}$$

Corollary X. $\lim_\mathrm{n} \mathrm{P}^\mathrm{n}$ *exists if and only if every positive recurrent component is aperiodic. In this case*

$$\lim_n P^n(x, y) = \begin{cases} 0, & y \notin \mathscr{L}_P \\ \pi_{C_i}(y), & x, y \in C_i \\ 0, & x \in C_i, y \in C_j \\ 0, & x \in \mathscr{L}_N, y \in C_i \\ \rho_{C_i}(x)\pi_{C_i}(y), & x \in \mathscr{L}_T, y \in C_i, \end{cases}$$

where $\mathscr{L}_P = \bigcup C_i$ *is the decomposition of* $\mathscr{L}_P$.

Corollary XI. *Let the periods* $d_1, d_2, \ldots$ *of the components* $C_1, C_2, \ldots$ *of* $\mathscr{L}_P$ *have l.c.m.* D. *Then each of the* D *limits*

$$\lim_n P^{nD+r}, \qquad 0 \le r < D,$$

exists and is distinct.

Geometric Ergodicity

An irreducible aperiodic positive recurrent Markov chain is said to be *geometrically ergodic* if for each pair of states $x, y \in \mathscr{L}$ there exist numbers $M(x, y) \ge 0$ and $\lambda(x, y) \in [0, 1)$ such that for any $n \ge 1$

$$|P^n(x, y) - \pi(y)| \le M(x, y)\lambda^n(x, y).$$

Theorem XII (from Parzen [45]). *A finite aperiodic irreducible Markov chain is geometrically ergodic.*

Proof. Set

$$M_n(y) = \max_x P^n(x, y), \qquad m_n(y) = \min_x P^n(x, y).$$

Since

$$P^{n+1}(x, y) = \sum_z P(x, z)P^n(z, y)$$

it follows that $\{M_n(y)\}$ is decreasing in n and $\{m_n(y)\}$ is increasing in n. Setting $d_n(y) = M_n(y) - m_n(y)$ we get that $\{d_n(y)\}$ is decreasing in n. Since the chain is finite and aperiodic irreducible we can find N such that $P^N(x, y) > 0$ for all $x, y \in \mathscr{L}$. Set $c = \min_{x,y} P^N(x, y)$. Then (assuming $|\mathscr{L}| > 1$) $0 < c \le \frac{1}{2}$. For any x, y

$$P^{(n+1)N}(x, y) = \sum_z P^N(x, z)P^{nN}(z, y).$$

Choose x, w such that

$$P^{(n+1)N}(x, y) = M_{(n+1)N}(y), \qquad P^{nN}(w, y) = m_{nN}(y).$$

Then

$$M_{(n+1)N}(y) = P^N(x, w)m_{nN}(y) + \sum_{z \neq w} P^N(x, z)P^{nN}(z, y)$$
$$\leq P^N(x, w)m_{nN}(y) + M_{nN}(y)[1 - P^N(x, w)]$$
$$\leq M_{nN}(y) - cd_{nN}(y).$$

Similarly by choosing x, w such that

$$P^{(n+1)N}(x, y) = m_{(n+1)N}(y), \qquad P^{nN}(w, y) = M_{nN}(y)$$

we obtain

$$m_{(n+1)N}(y) \geq m_{nN}(y) - cd_{nN}(y).$$

Combining these estimates leads to

$$d_{(n+1)N}(y) \leq (1 - 2c)d_{nN}(y)$$

and thus inductively

$$d_{nN}(y) \leq (1 - 2c)^{n-1}d_N(y).$$

Since $\{d_n(y)\}$ is decreasing in n we have

$$d_n(y) \leq (1 - 2c)^{\lfloor n/N \rfloor - 1}d_N(y) \leq (1 - 2c)^{(n/N)-2}d_N(y)$$

for $n \geq N$. Since $P^n(x, y)$ and $\pi(y)$ are both sandwiched in between $m_n(y)$ and $M_n(y)$ our result follows. □

Kendall has shown that an infinite irreducible aperiodic positive recurrent chain is geometrically ergodic if and only if the generating function $\mathbf{E}_y t^{T_y}$ has a radius of convergence greater than one, for some y.

Examples

We consider first the *two-state chain*. Its transition probability matrix is given by

$$P = \begin{pmatrix} q_1 & p_1 \\ p_2 & q_2 \end{pmatrix}$$

where $q_1 = 1 - p_1, q_2 = 1 - p_2$. If $p_1 + p_2 > 0$, then we have the explicit formula

$$P^n = \begin{bmatrix} \dfrac{p_2}{p_1 + p_2} & \dfrac{p_1}{p_1 + p_2} \\ \dfrac{p_2}{p_1 + p_2} & \dfrac{p_1}{p_1 + p_2} \end{bmatrix} + (1 - p_1 - p_2)^n \begin{bmatrix} \dfrac{p_1}{p_1 + p_2} & -\dfrac{p_1}{p_1 + p_2} \\ -\dfrac{p_2}{p_1 + p_2} & \dfrac{p_2}{p_1 + p_2} \end{bmatrix}.$$

The stationary distribution is given by

$$\pi = \left(\frac{p_2}{p_1 + p_2} \quad \frac{p_1}{p_1 + p_2}\right).$$

We distinguish three cases.

(A) If $p_1 = p_2 = 0$ then $\mathscr{L}_P$ consists of two closed irreducible sets $\{0\}$ and $\{1\}$. There are infinitely many stationary distributions. The one supported on $\{0\}$ is (1 0) and the one supported on $\{1\}$ is (0 1). For every n, $P^n = I$.

(B) If $0 < p_1 + p_2 < 2$ then the chain is aperiodic irreducible positive recurrent. There is a unique stationary distribution π and

$$\lim_n P^n = \begin{pmatrix} \pi(0) & \pi(1) \\ \pi(0) & \pi(1) \end{pmatrix}.$$

(C) If $p_1 = p_2 = 1$ then the chain is periodic with period 2. It is still irreducible positive recurrent, and there is still a unique stationary distribution, namely, $\pi = (\frac{1}{2}\ \frac{1}{2})$, but P^n does not have a limit. In fact for all n

$$P^{2n} = I, \qquad P^{2n+1} = P.$$

In any event

$$\lim_n \frac{1}{n} \sum_{m=1}^{n} P^m = \begin{pmatrix} \frac{1}{2} & \frac{1}{2} \\ \frac{1}{2} & \frac{1}{2} \end{pmatrix}.$$

The Ehrenfest chain with three balls has transition probability matrix

$$P = \begin{bmatrix} 0 & 1 & 0 & 0 \\ \frac{1}{3} & 0 & \frac{2}{3} & 0 \\ 0 & \frac{2}{3} & 0 & \frac{1}{3} \\ 0 & 0 & 1 & 0 \end{bmatrix}.$$

This chain has period $d = 2$ and stationary distribution $\pi = (\frac{1}{8}\ \frac{3}{8}\ \frac{3}{8}\ \frac{1}{8})$. The limiting behavior of P^n is

$$\lim_n P^{2n} = \begin{bmatrix} \frac{1}{4} & 0 & \frac{3}{4} & 0 \\ 0 & \frac{3}{4} & 0 & \frac{1}{4} \\ \frac{1}{4} & 0 & \frac{3}{4} & 0 \\ 0 & \frac{3}{4} & 0 & \frac{1}{4} \end{bmatrix}$$

$$\lim_n P^{2n+1} = \begin{bmatrix} 0 & \frac{3}{4} & 0 & \frac{1}{4} \\ \frac{1}{4} & 0 & \frac{3}{4} & 0 \\ 0 & \frac{3}{4} & 0 & \frac{1}{4} \\ \frac{1}{4} & 0 & \frac{3}{4} & 0 \end{bmatrix}.$$

We can modify the Ehrenfest chain, so as to "smooth" it, by returning the ball selected at random to a random urn. In this case the transition matrix is given by

$$P = \begin{bmatrix} \frac{1}{2} & \frac{1}{2} & 0 & 0 \\ \frac{1}{6} & \frac{1}{2} & \frac{1}{3} & 0 \\ 0 & \frac{1}{3} & \frac{1}{2} & \frac{1}{6} \\ 0 & 0 & \frac{1}{2} & \frac{1}{2} \end{bmatrix}.$$

The stationary distribution π is the same as earlier, $\pi = (\frac{1}{8}\,\frac{3}{8}\,\frac{3}{8}\,\frac{1}{8})$; but this time the period is $d = 1$, so

$$\lim_n P^n = \begin{pmatrix} \frac{1}{8} & \frac{3}{8} & \frac{3}{8} & \frac{1}{8} \\ \frac{1}{8} & \frac{3}{8} & \frac{3}{8} & \frac{1}{8} \\ \frac{1}{8} & \frac{3}{8} & \frac{3}{8} & \frac{1}{8} \\ \frac{1}{8} & \frac{3}{8} & \frac{3}{8} & \frac{1}{8} \end{pmatrix}.$$

Consider next a birth and death chain on $\mathscr{L} = \{0, 1, 2, \dots\}$ with transition probabilities q_x, r_x, p_x. Based on our computations of m_{xx} in Section IV we know that if the chain is irreducible then it is

$$\text{transient} \quad \Leftrightarrow \quad \sum_x \gamma_x < \infty$$

$$\text{positive recurrent} \quad \Leftrightarrow \quad \sum_x \frac{1}{p_x \gamma_x} < \infty$$

$$\text{null recurrent} \quad \Leftrightarrow \sum_x \gamma_x = \sum_x \frac{1}{p_x \gamma_x} = \infty.$$

Furthermore, in the positive recurrent case

$$\pi(x) = \frac{1}{m_{xx}} = \frac{1}{c p_x \gamma_x}, \tag{8}$$

where $c = \sum_x \frac{1}{p_x \gamma_x}$. If some $r_x > 0$, then the chain is aperiodic. Otherwise if $r_x = 0$ for all x then the period is 2. In this latter case

$$\lim_n P^{2n}(x, y) = \begin{cases} 0, & y - x \text{ odd} \\ 2\pi(y), & y - x \text{ even}; \end{cases}$$

$$\lim_n P^{2n+1}(x, y) = \begin{cases} 2\pi(y), & y - x \text{ odd} \\ 0, & y - x \text{ even}. \end{cases}$$

If $\mathscr{L}$ is finite and the chain is irreducible, then it is positive recurrent and (8) still holds.

Our next example is the $GI/D/1$ queueing chain from Section IV, with transition probabilities

$$P(x, y) = \begin{cases} p_Z(y), & x = 0 \\ p_Z(y - x + 1), & x \geq 1. \end{cases}$$

We showed earlier that $\mathbb{P}_x(T_{x-1} = n)$ is independent of x. Thus under $\mathbb{P}_x$ the times $T_0 - T_1, T_1 - T_2, \dots, T_{x-2} - T_{x-1}, T_{x-1}$ are i.i.d. We can think of T_0 then as the sum of x i.i.d. times $\sum_{k=1}^{x-1} (T_{k-1} - T_k) + T_{x-1}$. If G is the generating function of T_0, $G(t) = \mathbf{E}_1 t^{T_0}$ then

$$G(t) = \mathbf{E}_1 \mathbf{E}(t^{T_0} | X_1) = t\mathbf{E}_1 \mathbf{E}(t^{T_0 - 1} | X_1) = t\mathbf{E}G^{X_1}(t) = t\Phi_Z(G(t)).$$

Furthermore, since $P(1, y) = P(0, y)$, $y \geq 0$, the generating function G is also given by $G(t) = \mathbf{E}_0 t^{T_0}$. In order to decide when the chain is positive recurrent we need to compute $m_{00} = \mathbf{E}_0 T_0 = G'(1)$. Using the condition $G(t) = t\Phi_Z(G(t))$, we write

$$G'(t) = \frac{\Phi_Z(G(t))}{1 - t\Phi'_Z(G(t))},$$

and thus $\lim_{t \uparrow 1} G'(t) = \dfrac{1}{1-\mu}$. We conclude that the chain is positive recurrent if and only if $\mu < 1$. Since it is transient for $\mu > 1$ we see that the case $\mu = 1$ corresponds to null recurrence.

To find the stationary distribution observe that (from Parzen [45])

$$\pi(y) = \pi(0)p_Z(y) + \sum_{x=1}^{y+1} \pi(x)p_Z(y - x + 1).$$

Thus if B is the generating function

$$B(t) = \sum_{y=0}^{\infty} \pi(y)t^y,$$

then

$$B(t) = \pi(0)\Phi_Z(t) + \sum_{x=1}^{\infty} \pi(x)t^x\Phi_Z(t) = \Phi_Z(t)\left[\pi(0) + \frac{B(t) - \pi(0)}{t}\right],$$

and thus

$$B(t) = (1 - \mu)\frac{(t-1)\Phi_Z(t)}{t - \Phi_Z(t)}.$$

Recall that when $\mu < 1$ then Φ_Z has no fixed points in $[0, 1)$.

For our final example here, suppose that at times $n = 1, 2, \ldots, Z_n$ particles are added to a box, where $Z_1, Z_2, \ldots$ are independent Poisson random variables with intensity λ. Suppose that each particle in the box at time n, independently of all the other particles in the box, has probability $p < 1$ of remaining in the box at time $n + 1$, and probability $q = 1 - p$ of being removed from the box by time $n + 1$. Let X_n denote the number of particles in the box at time n. Then $\{X_n\}$ is a Markov chain. We are interested in its asymptotics.

By considering $X_{n+1} = Z_{n+1} + Y_n$, where Y_n is binomial with parameters X_n and p, we calculate the transition probabilities of $\{X_n\}$ to be

$$P(x, y) = \sum_{z=0}^{\min(x,y)} \binom{x}{z} p^z q^{x-z} e^{-\lambda} \frac{\lambda^{y-z}}{(y-z)!}.$$

We expect the stationary distribution to be Poisson with intensity λ/q; that is,

$$\pi(x) = e^{-\lambda/q}\frac{(\lambda/q)^x}{x!}, \qquad x \geq 0.$$

This can be checked directly, but of more significance here is that we can actually compute P^n explicitly.

To carry this out, observe that

$$\mathbf{E}X_{n+1} = \lambda + p\mathbf{E}X_n,$$

and so inductively

$$\mathbf{E}X_n = \frac{\lambda}{q}(1 - p^n) + p^n\mathbf{E}X_0.$$

If X_0 is Poisson distributed then X_n is too, since Y_n is a filtered Poisson. Thus for any $t > 0$ (taken to be the intensity of X_0)

$$\sum_{x=0}^{\infty} \frac{e^{-t}t^x}{x!} P^n(x, y) = \mathbb{P}(X_n = y)$$

$$= \exp\left\{-\left[\frac{\lambda}{q}(1 - p^n) + p^n t\right]\right\} \frac{\left[\frac{\lambda}{q}(1 - p^n) + p^n t\right]^y}{y!}.$$

Equating coefficients of like powers of t leads to

$$P^n(x, y) = e^{-(\lambda/q)(1-p^n)} \sum_{z=0}^{\min(x,y)} \binom{x}{z} p^{nz}(1 - p^n)^{x-z} \frac{\left[\frac{\lambda}{q}(1 - p^n)\right]^{y-z}}{(y - z)!}.$$

Since $p < 1$ we have $p^n \to 0$, and thus we verify directly that $\lim_n P^n(x, y) = \pi(y)$.

We pointed out in the beginning of this section that if $T_y^{(k)}$ denotes the time of the kth visit to y, then the waiting times between visits, $T_y^{(k)} - T_y^{(k-1)}$, $k \geq 2$, are i.i.d. Thus we can apply the Law of Large Numbers and the Central Limit Theorem to them. Doing so we find that for an irreducible positive recurrent chain

$$\lim_k \frac{T_y^{(k)}}{k} = m_{yy} \qquad \text{a.s.,} \tag{9}$$

and if $\mathbf{E}_y T_y^2 < \infty$ then

$$\frac{T_y^{(k)} - km_{yy}}{\sigma_{yy}\sqrt{k}} \xrightarrow{\mathscr{D}} N(0, 1),$$

where $\sigma_{yy}^2 = \text{Var}_y T_y$. Furthermore $N_n(y)$ is connected to the $T_y^{(k)}$s by

$$N_n(y) < k \Leftrightarrow T_y^{(k)} > n, \tag{10}$$

and by exploiting this we can also analyze the distribution of the occupation time $N_n(y)$ for large n.

Theorem XIII (from Parzen [45]). *For an irreducible positive recurrent chain*

$$\lim_n \frac{N_n(y)}{n} = \pi(y) \qquad \text{a.s.,} \tag{11}$$

$$\frac{N_n(y) - n\pi(y)}{\sigma_{yy}[\pi(y)]^{3/2}\sqrt{n}} \xrightarrow{\mathscr{D}} N(0, 1). \tag{12}$$

PROOF. We obtain (11) by dividing through by $N_n(y)$ in the inequality

$$T_y^{(N_n(y))} \leq n < T_y^{(N_n(y)+1)}$$

and using (9). (Note that $N_n(y) \to \infty$ as $n \to \infty$ since the chain is irreducible recurrent.) To obtain (12), observe that by (10)

$$\frac{N_n(y) - n\pi(y)}{\sigma_{yy}[\pi(y)]^{3/2}\sqrt{n}} \leq x \Leftrightarrow T_y^{(k)} > n$$

where k is the greatest integer

$$k = \lfloor \sigma_{yy}[\pi(y)]^{3/2}\sqrt{n}x + n\pi(y) \rfloor + 1.$$

As $n \to \infty$ we have

$$\frac{n - km_{yy}}{\sigma_{kk}\sqrt{k}} \to -x. \qquad \square$$

Observe that as a by-product of (12), we infer that $\lim_n \frac{1}{n} \operatorname{Var} N_n(y) = \sigma_{yy}^2 \pi^3(y)$.

As a final note on discrete Markov chains we point out that classifying an irreducible Markov chain on $\mathscr{L} = \{0, 1, 2, \ldots\}$ (transient, null recurrent, positive recurrent) can be formulated in terms of its *generator* $L = P - I$. This is the linear operator with action

$$Lu(x) = \sum_y P(x, y)u(y) - u(x).$$

Theorem XIV (from Parzen [45]). *Let* $\{X_n\}$ *be an irreducible Markov chain on* $\mathscr{L} = \{0, 1, 2, \ldots\}$ *with generator* $L = P - I$.

(i) *The chain is transient if and only if there is a nonconstant* $u \in l_\infty$ *with* $L_1 u = 0$, *where* L_1 *is the restriction of* L *to* $\{1, 2, \ldots\}$.

(ii) *The chain is positive recurrent if and only if there is a nonzero* $u \in l_1$ *with* $Lu = 0$.

(iii) *The chain is recurrent if there is a vector* u *satisfying* $\lim_x u(x) = \infty$ *with*

$$Lu(x) \leq 0, \qquad x \geq 1.$$

PROOF. (i) Convert 0 to an absorbing state by redefining $P(x, 0)$, $x \in \mathscr{L}$. We showed in Section IV that $L_1 u = 0$ has only the trivial solution in l_∞ if and only if $\rho_{x0} = 1$, $\forall x \geq 1$. Furthermore other than the zero solution $L_1 u = 0$ cannot have any constant solutions, since necessarily $P(x, 0) > 0$ for some $x \geq 1$. (Otherwise the original chain would not be irreducible.)

$\square$

(ii) This is just the existence of a stationary distribution. □

(iii) Again convert 0 to an absorbing state and denote the modified transition probabilities by $\hat{P}(x, y)$. Then

$$\sum_y \hat{P}(x, y)u(y) \le u(x), \qquad \forall x.$$

Iterating this we find that for any $n \ge 1$

$$\sum_y \hat{P}^n(x, y)u(y) \le u(x), \qquad \forall x. \tag{13}$$

Given $K > 0$ we can choose N so that $u(y) \ge K$ for all $y > N$. Then from (13)

$$\sum_{y=0}^{N} \hat{P}^n(x, y)u(y) + K\left[1 - \sum_{y=0}^{N} \hat{P}^n(x, y)\right] \le u(x).$$

The states $y = 1, 2, \ldots$ are all transient for the modified chain (since they lead to 0), and thus

$$\lim_n \hat{P}^n(x, y) = \begin{cases} \rho_{x0}, & y = 0, \\ 0, & y \ge 1. \end{cases}$$

Thus letting $n \to \infty$, we find that

$$1 - \rho_{x0} \le \frac{u(x) - \rho_{x0}u(0)}{K},$$

and since K was arbitrary it must be that $\rho_{x0} = 1$, $x \ge 1$. □

Exercises

1. (Hoel, Port and Stone [28], Parzen [45]) Regarding the matrices (a)–(l) in Problem 31,

   (i) find $\lim_{n\to\infty} \frac{1}{n} \sum_{m=1}^{n} P^m$.

   (ii) find $\lim_{n\to\infty} P^n(x, y)$ for each pair of states x, y for which this limit exists.

2. (Hoel, Port and Stone [28]) A particle moves according to a Markov chain on $\{1, 2, \ldots, c + d\}$. Starting from any one of the first c states, the particle jumps in one transition to a state chosen uniformly from the last d states; starting from any of the last d states, the particle jumps in one transition to a state chosen uniformly from the first c states. Find the stationary distribution.

3. (Parzen [45]) Suppose P is doubly stochastic; i.e.

   $$\sum_x P(x, y) = \sum_y P(x, y) = 1.$$

   Show that if the chain is infinite and irreducible then it cannot be positive recurrent.

4. (Hoel, Port and Stone [28]) $P(x, x+1) = p$, $P(x, 0) = 1 - p$ for $x \geq 0$. Find the stationary distribution.

5. (Hoel, Port and Stone [28]) Let π be a stationary distribution of a Markov chain.

   (i) Show that $\pi(x) > 0$, $x \to y \Rightarrow \pi(y) > 0$.
   (ii) Show that if

$$P(x, y) = cP(x, z), \qquad \forall x \in \mathscr{L}$$

   then $\pi(y) = c\pi(z)$.

6. (Parzen [45]) $P(x, 0) = \dfrac{x+1}{x+2}$, $P(x, x+1) = \dfrac{1}{x+2}$ for $x \geq 0$. Is this chain positive recurrent? If so, find its long-run distribution.
   What if $P(x, 0) = \dfrac{1}{x+2}$, $P(x, x+1) = \dfrac{x+1}{x+2}$?

7. (Parzen [45]) Show that if $x \leftrightarrow y$ then

$$P^m(x, y) > 0, \quad P^n(x, y) > 0 \Rightarrow d_y | n - m.$$

8. (Karlin and Taylor [31]) Let $1 = b_0 \geq b_1 \geq b_2 \geq \cdots$ be a decreasing sequence of positive numbers.

$$P(x, y) = \begin{cases} \dfrac{b_y}{b_x}(\beta_x - \beta_{x+1}), & y \leq x \\ \dfrac{\beta_{x+1}}{\beta_x}, & y = x + 1 \\ 0, & \text{elsewhere} \end{cases}$$

   where

$$\beta_x = \frac{b_x}{\sum_{i=0}^{x} b_i}.$$

   Find necessary and sufficient conditions for the chain to be transient, null recurrent, and positive recurrent. (Hint: compute $P^n(0, 0)$.)

9. (Karlin and Taylor [31]) Consider the mapping $F: R^{2k+1} \to \mathbb{R}^{2k+1}$

$$F(x)_i = \begin{cases} \tfrac{1}{2}(x_{i-1} + x_{i+1}), & 1 < i < 2k + 1 \\ \tfrac{1}{2}(x_{2k+1} + x_2), & i = 1 \\ \tfrac{1}{2}(x_{2k} + x_1), & i = 2k + 1. \end{cases}$$

   If $w = (w_1, \ldots, w_{2k+1})$, $w_i \geq 0$ find

$$\lim_{n \to \infty} F^n(w).$$

10. (Karlin and Taylor [31]) Let X_n, $n \geq 0$ be an irreducible chain with a finite state space. Let $\varphi(x)$ be a concave function, $x \geq 0$, and set

$$\alpha_k(y) = \sum_{x \in \mathscr{L}} \pi(x)\varphi(P^k(x, y)).$$

   Show that

$$\alpha_0(y) \leq \alpha_1(y) \leq \alpha_2(y) \leq \cdots.$$

11. (Hoel, Port and Stone [28]) Let X_n be a positive recurrent birth and death chain, and suppose X_0 has the stationary distribution π. Find

$$P(X_0 = y | X_1 = x)$$

12.

$$P = \begin{bmatrix} 0 & 1 & 0 & 0 & 0 & \cdots \\ q & 0 & p & 0 & 0 & \cdots \\ 0 & q & 0 & p & 0 & \cdots \\ 0 & 0 & q & 0 & p & \cdots \\ \cdots & \cdots & \cdots & \cdots & \cdots & \cdots \end{bmatrix}$$

Find π.

13. Find the stationary distribution of the Ehrenfest chain. Find the mean and variance of this distribution.

14. (Hoel, Port and Stone [28]) Let X_n, $n \geq 0$ be as in our example on pages 114 and 115. If X_0 has the stationary distribution find

$$\mathrm{Cov}(X_m, X_{m+n}).$$

15.

$$P(x, y) = \begin{cases} p, & y = x + 1 \\ q, & y = 0 \\ 0, & \text{otherwise.} \end{cases}$$

Find π.

16.

$$P(x, y) = \begin{cases} p_y, & x = 0 \\ \dfrac{1}{x}, & x \geq 1, 0 \leq y < x. \\ 0, & x \geq 1, y \geq x. \end{cases}$$

Assume $p_y > 0$ ($y = 0, 1, 2, \ldots$). Find necessary and sufficient conditions on $\{p_y\}_{y=0}^{\infty}$ for the chain to be transient/null recurrent/positive recurrent. In this last case, find π.

17. (Hoel, Port and Stone [28])

$$P = \begin{bmatrix} 0 & \frac{1}{3} & \frac{2}{3} & 0 & 0 \\ 0 & 0 & 0 & \frac{1}{4} & \frac{3}{4} \\ 0 & 0 & 0 & \frac{1}{4} & \frac{3}{4} \\ 1 & 0 & 0 & 0 & 0 \\ 1 & 0 & 0 & 0 & 0 \end{bmatrix}$$

Find the period and stationary distribution.

18. (Hoel, Port and Stone [28]) Let X_n, $n \geq 0$ be the Ehrenfest chain with $d = 4$. Analyze P^n for large n.

19. (Karlin and Taylor [31]) Suppose that the weather on any day depends on the weather conditions for the previous two days. We say the state is

(S, S)—sunny both today and yesterday,
(S, C)—sunny yesterday, cloudy today,
(C, S)—cloudy yesterday, sunny today,
(C, C)—cloudy both today and yesterday.

	(S, S)	(S, C)	(C, S)	(C, C)
(S, S)	0.8	0.2	0	0
(S, C)	0	0	0.4	0.6
(C, S)	0.6	0.4	0	0
(C, C)	0	0	0.1	0.9

On what fraction of days in the long run is it sunny?

20. (Karlin and Taylor [31]) Let $X_1, X_2, \ldots$ be i.i.d. with

$$P(X_k = +1) = p, \qquad P(X_k = -1) = q = 1 - p.$$

Assume $p \geq q$.
Set $S_0 = 0$.

$$S_n = X_1 + \cdots + X_n, \qquad n \geq 1$$

$$M_n = \max_{0 \leq k \leq n} S_k, \qquad Y_n = M_n - S_n$$

$$T(s) = \min(n: S_n = s) \qquad \tau(y) = \min(n: Y_n = y)$$

(a) Prove

$$P\left(\max_{0 \leq k \leq T(s)} Y_k < y\right) = \begin{cases} \left[\dfrac{\dfrac{p}{q} - \left(\dfrac{p}{q}\right)^{y+1}}{1 - \left(\dfrac{p}{q}\right)^{y+1}}\right]^s, & p \neq q \\ \left(\dfrac{y}{1+y}\right)^s, & p = q \end{cases}$$

(b) Prove $$P(M_{\tau(y)} \geq a) = \theta^a.$$

Find θ as well here.
Hint: (M_n, Y_n) is a random walk on $\mathbb{Z}^2$.

Section VI
Markov Jump Processes*

Pure Jump Processes

We want to describe Markov processes that evolve through continuous time $t \geq 0$, but in a discrete state space $\mathscr{S}$. The prescription for such a process has two ingredients. There are random *jump times* $0 < \tau_1 < \tau_2 < \cdots < \tau_n < \cdots$ when the process jumps away from the state it is at, and there are *transition probabilities* Q_{xy} that govern the transitions at these jump times. The process $\{X(t): t \geq 0\}$ itself has piecewise constant paths, which we can take to be right-continuous

$$X(t) = \begin{cases} x_0, & 0 \leq t < \tau_1, \\ x_1, & \tau_1 \leq t < \tau_2, \\ x_2, & \tau_2 \leq t < \tau_3, \\ \cdots\cdots\cdots\cdots\cdots\cdots\cdots & \end{cases}$$

The jump times are described in terms of the *waiting time* d.f.s $F_x(t)$, which correspond to the time the process waits at state x until it jumps away. The jump process is *pure* or *nonexplosive if* $\mathbb{P}_x\left(\lim_n \tau_n = \infty\right) = 1$ for every $x \in \mathscr{S}$. Otherwise it is *explosive*. If $\mathscr{S}$ is finite, then the process is necessarily nonexplosive. The transition probabilities satisfy

$$Q_{xx} = 0, \qquad \sum_y Q_{xy} = 1,$$

unless x is an absorbing state, in which case $Q_{xy} = \delta_{xy}$.

There are two fundamental stipulations we make if the process is to be Markovian.

* Material taken from Hoel, Port and Stone [28] with permission.

(A) **Independence of transitions and waiting times.** *For any* x, y $\in \mathscr{L}$ *and* t ≥ 0

$$\mathbb{P}_x(\tau_1 \leq t, X(\tau_1) = y) = F_x(t)Q_{xy}.$$

(B) **Renewal assumption.** *Whenever and however the process gets to be at a state* y, *it acts just as a process starting initially at* y.

An example of the use of these assumptions is

$$\mathbb{P}_x(\tau_1 \leq s, X(\tau_1) = y, \tau_2 - \tau_1 \leq t, X(\tau_2) = z) = F_x(s)Q_{xy}F_y(t)Q_{yz}.$$

One important consequence of the renewal assumption is that for any $x \in \mathscr{L}$ and $s, t \geq 0$

$$\mathbb{P}_x(\tau_1 > t + s | \tau_1 > s) = \mathbb{P}_x(\tau_1 > t). \tag{1}$$

This condition implies (by Cauchy's Lemma in Section III) that for a non-absorbing state x the waiting time at x must be exponentially distributed; that is, $F_x(t) = 1 - e^{-q_x t}$, $t \geq 0$, for some $q_x > 0$. The parameter q_x is then the *rate* at which jumps occur, $\frac{1}{q_x}$ being the *mean waiting time* at x. If x is absorbing, then we set $q_x = 0$.

A quantity we will need to compute is the discrete density of $X(t)$,

$$P_{xy}(t) = \mathbb{P}_x(X(t) = y).$$

It follows from assumptions (A) and (B) that, in analogy to the discrete time chains, our Markov process $\{X(t): t \geq 0\}$ satisfies the *Markov property*,

$$\mathbb{P}(X(t) = y | X(s_1) = x_1, \ldots, X(s_n) = x_n, X(s) = x) = P_{xy}(t - s)$$

for states $x_1, \ldots, x_n, x, y \in \mathscr{L}$ and times $0 \leq s_1 < s_2 < \cdots < s_n < s < t$. Thus the probabilities $P_{xy}(t)$ convey full distributional information about the entire process $\{X(t): t \geq 0\}$. That is, they convey all joint finite-dimensional distributions $F_{X(t_1), \ldots, X(t_n)}$ for any $n \geq 1$ and any times $0 \leq t_1 < \cdots < t_n$. We can think of the probabilities $P_{xy}(t)$ as entries in a (finite or countably infinite) matrix $P(t) = (P_{xy}(t))_{x, y \in \mathscr{L}}$. If $\pi(0) = (\pi_x(0))_{x \in \mathscr{L}}$ is the initial distribution for $X(0)$, then, as in the discrete time case, the distribution $\pi(t)$ for $X(t)$ is given by

$$\pi(t) = \pi(0)P(t).$$

Here we are treating π as a (finite or countably infinite) row vector. This tells us that the natural setting for $P(t)$ is that of an operator from l_1 to l_1. With respect to these spaces, $P(t)$ is a contraction, and in fact $\|uP(t)\|_1 = \|u\|_1$ whenever $u \in l_1$ is nonnegative. Observe that entrywise $\lim_{t \downarrow 0} P(t) = I$, the (finite or countably infinite) identity matrix; that is,

$$\lim_{t \downarrow 0} P_{xy}(t) = \delta_{xy}, \qquad x, y \in \mathscr{L}. \tag{2}$$

Indeed for $x \neq y$

$$\mathbb{P}_x(X(t) = y) \le \mathbb{P}_x(\tau_1 \le t)$$

and

$$\mathbb{P}_x(X(t) = x) \ge \mathbb{P}_x(\tau_1 > t).$$

From this follows, using the Dominated Convergence Theorem that for any $u \in l_1$

$$\lim_{t \downarrow 0} \|uP(t) - u\|_1 = 0. \tag{3}$$

The probability densities $P_{xy}(t)$ satisfy

$$P_{xy}(t + s) = \sum_z P_{xz}(t)P_{zy}(s),$$

or in terms of the operators $P(t)$

$$P(t + s) = P(t)P(s). \tag{4}$$

This equation is called the *Chapman-Kolmogorov equation*, and it has significant implications about the family $\{P(t)\colon t \ge 0\}$. It implies that this family forms a one-parameter semigroup. Thus from (3) we conclude that $\{P(t)\colon t \ge 0\}$ is a strongly continuous contraction semigroup on l_1. Observe that for the operator norm

$$\|P(t) - I\|_1 = 2 \sup_x \,[1 - P_{xx}(t)],$$

so that this semigroup is uniformly continuous if and only if the rates q_x are bounded. In any event, we are led to inquire after the generator.

Theorem I. *The generator* $\mathrm{q} = (\mathrm{q}_{xy})$ *is given by*

$$q_{xy} = \begin{cases} -q_x, & y = x, \\ q_x Q_{xy}, & y \neq x. \end{cases}$$

That is, $\mathrm{q} = \mathrm{diag}(\mathrm{q}_x)(\mathrm{Q} - \mathrm{I})$.

Proof. Observe first that

$$\begin{aligned} \mathbb{P}_x(\tau_1 \le t, X(t) = y) &= \sum_{z \neq x} \mathbb{P}_x(\tau_1 \le t, X(\tau_1) = z, X(t) = y) \\ &= \sum_{z \neq x} \int_0^t q_x e^{-q_x s} Q_{xz} P_{zy}(t - s)\, ds \\ &= q_x e^{-q_x t} \int_0^t e^{q_x s} \sum_{z \neq x} Q_{xz} P_{zy}(s)\, ds. \end{aligned}$$

Since

$$\mathbb{P}_x(\tau_1 > t, X(t) = y) = \delta_{xy} e^{-q_x t},$$

we can combine these probabilities to arrive at

$$P_{xy}(t) = \delta_{xy} e^{-q_x t} + q_x e^{-q_x t} \int_0^t e^{q_x s} \sum_{z \neq x} Q_{xz} P_{zy}(s)\, ds.$$

Now differentiate and set $t = 0$. □

The generator q here corresponds to $P - I$ in the discrete time case. The entries q_{xy} are called the *infinitesimal parameters*, since for small t

$$P_{xy}(t) = \begin{cases} q_{xy} t + o(t), & y \neq x, \\ 1 - q_x t + o(t), & y = x. \end{cases}$$

Observe that $\sum_y q_{xy} = 0$, $\forall x$, and that $\sum_y |q_{xy}| = 2q_x$, so that q is bounded if and only if the rates q_x are bounded. The semigroup satisfies $\frac{dP}{dt} = qP = Pq$. The former equation is called the *backward equation*, and written out in component form it becomes

$$P'_{xy}(t) = \sum_z q_{xz} P_{zy}(t).$$

The latter equation is called the *forward equation* and written out it becomes

$$P'_{xy}(t) = \sum_z P_{xz}(t) q_{zy}.$$

Each of these equations has its merits, as we shall see in the examples that follow.

By working in the dual space we can also treat the operators $\{P(t): t \geq 0\}$ as a one-parameter contraction semigroup on l_∞, with action $u \mapsto P(t)u$, handling u as a (finite or countably infinite) column vector. The probabilistic meaning is that

$$(P(t)u)_x = \mathbf{E}_x u_{X(t)}.$$

The operators $P(t)$ are positivity-preserving, so that $P(t)u \geq 0$ whenever $u \geq 0$; and they satisfy $P(t)1 = 1$, where 1 denotes the constant vector of ones. Such a semigroup is called a *Markov semigroup*. The difficulty with this approach is that (2) no longer suffices to ensure strong continuity at $t = 0$, since l_1 is not reflexive. (Otherwise we could use the "weak equals strong" principle for semigroups. See Pazy [46].) For discussions of jump processes and semigroups, refer to Doob [13], Ethier, and Kurtz [16]; Feller [17]; and Hille and Phillips [26].

We can embed our continuous time process into a discrete time chain by defining $\hat{X}_n = X(\tau_n)$. This chain is called the *embedded chain* and its transition probabilities are Q_{xy}. In doing this embedding we lose track of real time, but passage phenomena do not change. As in the discrete time setting we define the *hitting time* $T_y = \min(t \geq \tau_1: X(t) = y)$ and *passage probability* $\rho_{xy} = \mathbb{P}_x(T_y < \infty)$. We say that a state y is *recurrent* if $\rho_{yy} = 1$ and *transient* otherwise, and we say that the process is *irreducible* if $\rho_{xy} > 0$ for all $x, y \in \mathscr{L}$.

Theorem II. *The passage probabilities $\hat{\rho}_{xy}$ for the embedded chain are equal to the corresponding probabilities ρ_{xy} for the jump process. In particular, we have the same decomposition $\mathscr{L}_R$ and $\mathscr{L}_T$, and an irreducible jump process is either recurrent or transient, according to the nature of the embedded chain.*

PROOF. Use the fact that

$$\hat{T}_y = n \Leftrightarrow T_y = \tau_n. \qquad \square$$

A discrete density $\pi = (\pi_x)_{x \in \mathscr{L}}$ is *stationary* for the jump process $\{X(t)\}$ if $\pi P(t) = \pi$ for all $t \geq 0$. This is equivalent to the condition

$$\pi q = 0, \tag{5}$$

since for any $t > 0$

$$\pi P(t) - \pi = \int_0^t \pi q P(s)\, ds.$$

A state y is *positive recurrent* if $m_{yy} = \mathbf{E}_y T_y < \infty$, and *null recurrent* if $\rho_{yy} = 1$ but $m_{yy} = \infty$. An irreducible recurrent process must be either null or positive recurrent. In the latter case the unique stationary distribution is given by

$$\pi_x = \frac{1}{q_x m_{xx}} = \frac{c}{q_x \hat{m}_{xx}},$$

since (π_x) is stationary for $\{X(t)\}$ if and only if $(q_x \pi_x)$ is in the null space of $Q - I$. Markov pure jump processes do not have periodicities. Thus an irreducible positive recurrent process behaves like an aperiodic chain, namely,

$$\lim_{t \uparrow \infty} P_{xy}(t) = \pi(y), \qquad x, y \in \mathscr{L}.$$

Poisson Process

A jump process $\{X(t)\}$ is said to be a *birth and death process* if $q_{xy} = 0$ whenever $|x - y| > 1$. Let us denote $\lambda_x = q_{x\,x+1}$ and $\mu_x = q_{x\,x-1}$. We refer to these parameters as the *birth and death rates*, respectively. In terms of them, the jump rates and transition probabilities are given by

$$q_x = \lambda_x + \mu_x, \qquad Q_{x\,x+1} = \frac{\lambda_x}{\lambda_x + \mu_x}, \qquad Q_{x\,x-1} = \frac{\mu_x}{\lambda_x + \mu_x}.$$

The process is a *pure birth process* if $\mu_x \equiv 0$ and a *pure death process* if $\lambda_x \equiv 0$. For a birth and death process the backward and forward equations become, respectively,

$$P'_{xy}(t) = \mu_x P_{x-1\,y}(t) - (\lambda_x + \mu_x)P_{xy}(t) + \lambda_x P_{x\,x+1}(t), \tag{6}$$

$$P'_{xy}(t) = \lambda_{y-1} P_{x\,y-1}(t) - (\lambda_y + \mu_y)P_{xy}(t) + \mu_{y+1} P_{x\,y+1}(t). \tag{7}$$

For a pure birth process we can solve the forward equation recursively,

$$P_{xy}(t) = \begin{cases} e^{-\lambda_x t}, & y = x, \\ \lambda_{y-1} \displaystyle\int_0^t e^{-\lambda_y(t-s)} P_{x\,y-1}(s)\,ds, & y > x, \\ 0, & y < x. \end{cases} \tag{8}$$

Similarly, for a pure death process we can solve the backward equation recursively.

Theorem III (from Feller [17]). *A necessary and sufficient condition for a pure birth process to be nonexplosive is that* $\sum_x \lambda_x^{-1}$ *diverges.*

PROOF. For $z \geq x$ set $S_z(t) = \mathbb{P}_x(X(t) \leq z)$. Then $S_z(t) = \sum_{y=x}^{z} P_{xy}(t)$ and it follows from the forward equation that

$$S'_z(t) = -\lambda_z P_{xz}(t).$$

Integrating, we obtain

$$\mathbb{P}_x(X(t) = \infty) \leq 1 - S_z(t) = \lambda_z \int_0^t P_{xz}(s)\,ds \leq 1.$$

Summing over $z = 1, \ldots, n$ we arrive at the inequality

$$\mathbb{P}_x(X(t) = \infty) \sum_{z=1}^{n} \lambda_z^{-1} \leq \int_0^t S_n(s)\,ds \leq \sum_{z=1}^{n} \lambda_z^{-1}.$$

If $\sum_{z=1}^{\infty} \lambda_z^{-1} = \infty$ then by the left inequality $\mathbb{P}_x(X(t) = \infty) = 0$ for all t. If $\sum_{z=1}^{\infty} \lambda_z^{-1} < \infty$ then for some t we must have $\lim_n S_n(t) < 1$; otherwise by the Dominated Convergence Theorem $\lim_n \int_0^t S_n(s)\,ds = t$, and this cannot be bounded uniformly in t. □

For a general birth and death process it can be shown that a sufficient condition for no explosion is that there exist constants A and B such that $\lambda_x \leq A + Bx$ for all $x \in \mathscr{L}$.

The *Poisson process* with rate $\lambda > 0$ is a pure birth process with constant birth rate $\lambda_x \equiv \lambda$. In this case the solution of the forward equation works out to be

$$P_{xy}(t) = \frac{(\lambda t)^{y-x}}{(y-x)!} e^{-\lambda t}, \qquad 0 \leq x \leq y.$$

Thus, under $\mathbb{P}_x$ the distribution of $X(t) - x$ is Poisson with parameter λt. Observe that the Poisson process is not only temporally homogeneous, like all of the jump processes we have been considering, but it is also *spatially homogeneous*. That is,

$$P_{xy}(t) = P_{0\,y-x}(t). \tag{9}$$

The two most important properties of the Poisson process are that it has stationary and independent increments, which we will describe now for a general stochastic process.

A continuous parameter stochastic process $\{X(t): t \geq 0\}$ is said to have *independent increments* if for all choices of n and $t_0 < t_1 < \cdots < t_n$ the n random variables $X(t_k) - X(t_{k-1})$, $1 \leq k \leq n$ are independent. The process is said to have *stationary increments* if for any $t_1 < t_2$ and $h > 0$ the random variables $X(t_2) - X(t_1)$ and $X(t_2 + h) - X(t_1 + h)$ have the same distribution. Observe that for a process with stationary independent increments, knowledge of the distribution of $X(t)$ for $t \in (0, \varepsilon)$ is enough to deduce the joint distribution of any n random variables $X(t_1)$, $\ldots$, $X(t_n)$ since

$$\varphi_{X(t_1),\ldots,X(t_n)}(u_1, \ldots, u_n) = \prod_{k=1}^{n} \varphi_{X(t_k - t_{k-1})}\left(\sum_{j=k}^{n} u_j\right).$$

(Take $t_0 = 0$ here.) In fact it can be shown that knowledge of the distribution of $X(t)$ at a *single* time $t > 0$ suffices to infer all joint distributions.

To see that indeed the Poisson process is a stationary independent increment process observe that if $t_0 < t_1 < \cdots < t_n$ then by (9)

$$\begin{aligned}
\mathbb{P}(X(t_1) - X(t_0) &= z_0, \ldots, X(t_n) - X(t_{n-1}) = z_n) \\
&= \sum_x \mathbb{P}(X(t_0) = x) P_{0z_0}(t_1 - t_0) \ldots P_{0z_n}(t_n - t_{n-1}) \\
&= P_{0z_0}(t_1 - t_0) \ldots P_{0z_n}(t_n - t_{n-1}).
\end{aligned}$$

Similarly if $t_1 < t_2$ then

$$\begin{aligned}
\mathbb{P}(X(t_2) - X(t_1) = z) &= \sum_x \mathbb{P}(X(t_1) = x, X(t_2) = x + z) \\
&= \sum_x \mathbb{P}(X(t_1) = x) P_{0z}(t_2 - t_1) = P_{0z}(t_2 - t_1).
\end{aligned}$$

The Poisson process can be used to model events occurring in time, such as calls coming into a telephone exchange, customers arriving at a queue, and radioactive disintegrations. In this framework $X(t)$ denotes the number of events occurring in the time interval $(0, t]$. For $s \leq t$ the random variable $X(t) - X(s)$ denotes the number of events in the time interval $(s, t]$. The waiting times between successive events are independent and exponentially distributed with common parameter λ if and only if $\{X(t)\}$ is a Poisson process. Observe that the time τ_n of the nth jump equals the hitting time T_n when the process reaches state n and the nth event occurs, assuming $X(0) = 0$. Furthermore, this time has a gamma distribution with parameters n and λ.

Theorem IV. *Let* $\{\mathrm{X(t)}\}$ *be a Poisson process with* $\mathrm{X(0) = 0}$. *Then conditioned on* $\mathrm{X(t) = k}$ *the times when the first* k *events occur are independent* $\mathrm{U(0, t)}$. *In particular for* $\mathrm{s \leq t}$ *the distribution of* $\mathrm{X(s)}$ *conditioned on* $\mathrm{X(t) = k}$ *is binomial with parameters* k *and* $\frac{\mathrm{s}}{\mathrm{t}}$.

PROOF. Consider an arbitrary partition $0 = t_0 < t_1 < \cdots < t_n = t$ and set $X_i = X(t_i) - X(t_{i-1})$, $1 \leq i \leq n$. Then if $\sum_i x_i = k$,

$$\mathbb{P}(X_1 = x_1, \ldots, X_n = x_n | X(t) = k) = \binom{k}{x_1, \ldots, x_n} \prod_{i=1}^{n} \left(\frac{t_i - t_{i-1}}{t}\right)^{x_i}. \tag{10}$$

Let $\xi_1, \ldots, \xi_k$ be the times of the first k events, ordered independently of $\{X(t)\}$, so that each ξ_i is equally likely to be any of $\tau_1, \ldots, \tau_k$. We want to show that for any Borel subset $B \subseteq [0, t]^k$,

$$\mathbb{P}((\xi_1, \ldots, \xi_k) \in B) = \frac{\mathrm{vol}(B)}{t^k}. \tag{11}$$

It follows from (10) that (11) holds when B is a product set

$$B = I_1 \times \cdots \times I_k,$$

and any pair of sets I_i and I_j are either identical or disjoint. Since any product set can be partitioned into such special product sets, and since both sides of (11) are additive in B, this suffices to conclude that (11) holds for all Borel subsets $B \subseteq \mathbb{R}^k$. □

There are several stochastic processes constructed by modifying or generalizing the Poisson process. The first one we describe is the *filtered Poisson process*. In this model events occur according to a Poisson process $\{X(t)\}$ with rate λ, but each event is independently recorded only with a probability p. The filtered process $\{Y(t)\}$ is easily seen to be a Poisson process with rate λp. Indeed $\{Y(t)\}$ has independent increments and for $s < t$ the conditional distribution of $Y(t) - Y(s)$ given $X(t) - X(s)$ is binomial with parameters $X(t) - X(s)$ and p.

Another modified process is the *compound Poisson process*. Here the rate λ is itself a nonnegative random variable Λ. In this case

$$\varphi_{X(t)}(u) = \varphi_\Lambda\left(\frac{e^{iu} - 1}{u} t\right).$$

A different sort of generalization is the Poisson point process. Here, instead of considering events as occurring in time, we think of points as being distributed randomly in space. Thus, consider an array of points distributed in a space $S \subseteq \mathbb{R}^m$. For each Borel subset $B \subseteq S$ let $N(B)$ denote the number of points (finite or infinite) contained in B. The array of points is said to form a *Poisson point process with intensity* λ if:

(i) *The number of points in nonoverlapping regions are independent random variables; i.e. if* $B_1, B_2, \ldots, B_n$ *are disjoint regions then* $N(B_1)$, $N(B_2), \ldots, N(B_n)$ *are independent.*
(ii) *For any region* B *of finite volume* vol(B), N(B) *is Poisson-distributed with rate* λ vol(B).

Notice how these assumptions are analogous to the properties of independent and stationary increments, respectively.

In order to enable us to study queues we close here with a useful property of exponential holding times.

Lemma V. *Let* $\xi_1, \ldots, \xi_n$ *be independent exponential holding times with respective parameters* $\alpha_1, \ldots, \alpha_n$. *Then* $\min(\xi_1, \ldots, \xi_n)$ *is exponential with parameter* $\alpha_1 + \cdots + \alpha_n$, *and*

$$\mathbb{P}(\xi_k = \min(\xi_1, \ldots, \xi_n)) = \frac{\alpha_k}{\alpha_1 + \cdots + \alpha_n}.$$

Moreover, with probability one $\xi_1, \ldots, \xi_n$ *take on* n *distinct values.*

PROOF.

$$\mathbb{P}(\min(\xi_1, \ldots, \xi_n) > t) = \mathbb{P}(\xi_1 > t) \ldots \mathbb{P}(\xi_n > t) = e^{-(\alpha_1 + \cdots + \alpha_n)t}$$

Set $\eta_k = \min(\xi_j\colon j \neq k)$. Then η_k is exponential with parameter $\beta_k = \sum_{j \neq k} \alpha_j$, and ξ_k, η_k are independent. Thus

$$\mathbb{P}(\xi_k = \min(\xi_1, \ldots, \xi_n)) = \mathbb{P}(\xi_k \leq \eta_k)$$
$$= \int_0^\infty \int_x^\infty \alpha_k e^{-\alpha_k x} \beta_k e^{-\beta_k y}\, dy\, dx = \frac{\alpha_k}{\alpha_k + \beta_k}.$$

The last statement in the lemma follows from the fact that $\xi_1, \ldots, \xi_n$ are joint absolutely continuous and the set where they do not assume distinct values is of (Lebesgue) measure zero. □

Birth and Death Processes

Let $\{X(t)\}$ be an irreducible birth and death process with birth and death rates λ_x and μ_x, respectively. Define

$$\gamma_y = \frac{\mu_1 \ldots \mu_y}{\lambda_1 \ldots \lambda_y}, \qquad y \in \mathscr{L}.$$

Make the assumption that

$$\lambda_x \leq A + Bx, \qquad x \in \mathscr{L},$$

for some constants A and B, so that we can be sure our process is non-explosive. Based on our results for the embedded (birth and death) chain we know that the birth and death process is transient if and only if $\sum_y \gamma_y < \infty$. By considering the stationary condition $\pi q = 0$ we find that the process is positive recurrent if and only if $\sum_y \frac{1}{\lambda_y \gamma_y} < \infty$, in which case the stationary distribution is given by

$$\pi(x) = \frac{c}{\lambda_x \gamma_x}, \qquad x \in \mathscr{L}, \tag{12}$$

where $c = \left(\sum_y \frac{1}{\lambda_y \gamma_y}\right)^{-1}$. Here we take $\gamma_0 = 1$. Formula (12) holds for a finite state process, $\mathscr{L} = \{0, 1, \ldots, d\}$, as well as a countably infinite state process, $\mathscr{L} = \{0, 1, \ldots\}$.

Our first example is the *two-state process*, whose generator is given by

$$q = \begin{pmatrix} -\lambda & \lambda \\ \mu & -\mu \end{pmatrix}.$$

One can integrate the forward or backward equation to arrive at

$$P(t) = \begin{bmatrix} \frac{\mu}{\lambda + \mu} & \frac{\lambda}{\lambda + \mu} \\ \frac{\mu}{\lambda + \mu} & \frac{\lambda}{\lambda + \mu} \end{bmatrix} + e^{-(\lambda+\mu)t} \begin{bmatrix} \frac{\lambda}{\lambda + \mu} & -\frac{\lambda}{\lambda + \mu} \\ -\frac{\mu}{\lambda + \mu} & \frac{\mu}{\lambda + \mu} \end{bmatrix}$$

For any given initial distribution $\pi(0) = (\pi_0 \quad \pi_1)$ we have

$$\pi(t) = \left(\frac{\mu}{\lambda + \mu} \quad \frac{\lambda}{\lambda + \mu}\right) + e^{-(\lambda+\mu)t}\left(\pi_0 - \frac{\mu}{\lambda + \mu} \quad \pi_1 - \frac{\lambda}{\lambda + \mu}\right).$$

The stationary distribution is $\pi = \left(\frac{\mu}{\lambda + \mu} \quad \frac{\lambda}{\lambda + \mu}\right)$.

Our next example (from Parzen [45]) is the *N-server queue*. We shall model the $M/M/N$ queue. Customers arrive according to a Poisson process with rate $\lambda > 0$. They are served by N servers, and the service times are exponentially distributed with rate μ. Whenever there are more than N customers waiting for service the excess customers form a queue. Let $X(t)$ denote the total number of customers waiting for service or being served at time t. Then $\{X(t)\}$ is a jump process with parameters $q_x = \lambda + \mu \min(x, N)$, $Q_{x\,x+1} = \frac{\lambda}{q_x}$ and $Q_{x\,x-1} = \frac{\mu}{q_x}$. Thus the birth and death rates are $\lambda_x \equiv \lambda$, $\mu_x = \mu \min(x, N)$. The terms γ_y are given by

$$\gamma_y = \begin{cases} y! \left(\dfrac{\mu}{\lambda}\right)^y, & y < N, \\ N!\, N^{y-N} \left(\dfrac{\mu}{\lambda}\right)^y, & y \geq N. \end{cases}$$

For large y we have $\gamma_y \sim \left(\dfrac{N\mu}{\lambda}\right)^y$, and so the process is transient if $N\mu < \lambda$, null recurrent if $N\mu = \lambda$, and positive recurrent if $N\mu > \lambda$. In this latter case the stationary distribution is given by

$$\pi(x) = \begin{cases} c \dfrac{\left(\dfrac{\lambda}{\mu}\right)^x}{x!}, & x < N, \\ c \dfrac{\left(\dfrac{\lambda}{\mu}\right)^x}{N!\, N^{x-N}}, & x \geq N, \end{cases}$$

where

$$c^{-1} = \sum_{x=0}^{N-1} \frac{\left(\dfrac{\lambda}{\mu}\right)^x}{x!} + \frac{N\mu}{N\mu - \lambda} \frac{\left(\dfrac{\lambda}{\mu}\right)^N}{N!}.$$

Observe that if $N = 1$ then π is geometric.

We are interested in the stationary distribution of $W(t)$, the time a customer arriving right after time t must wait to *begin* service. We assume that the queue is operating at its stationary distribution π. The waiting time distribution for $W(t)$ will be a mixed distribution, with an atom at $w = 0$, and absolutely continuous for $w > 0$. The mass at $w = 0$ is easy to evaluate, since

$$\mathbb{P}(W(t) > 0) = \mathbb{P}(X(t) \geq N) = \sum_{x=N}^{\infty} \pi(x) = \frac{N\mu}{N\mu - \lambda} \pi(N).$$

If $X(t) = N + n$ for some $n \geq 0$ the waiting $W(t)$ is gamma-distributed with parameters $n + 1$ and $N\mu$. Thus

$$\mathbb{P}(W(t) \geq w \mid X(t) = N + n) = \int_w^{\infty} N\mu \frac{(N\mu s)^n}{n!} e^{-N\mu s}\, ds.$$

Since

$$\mathbb{P}(X(t) = N + n) = \pi(N) \left(\frac{\lambda}{N\mu}\right)^n$$

we have

$$\mathbb{P}(W(t) \geq w) = \sum_{n=0}^{\infty} \int_w^{\infty} N\mu\pi(N) \frac{(\lambda s)^n}{n!} e^{-N\mu s}\, ds = \pi(N) \frac{N\mu}{N\mu - \lambda} e^{-(N\mu - \lambda)w}.$$

We define the *utilization factor*

$$\rho = \frac{\lambda}{N\mu}.$$

In terms of ρ the distribution of $W(t)$ is given by

$$\mathbb{P}(W(t) > 0) = \frac{\pi(N)}{1-\rho}$$

$$\mathbb{P}(W(t) \geq w) = \frac{\pi(N)}{1-\rho} e^{-(1-\rho)N\mu w}.$$

We measure the efficiency of our queue by a *customer loss ratio R*, defined to be

$$R = \mu EW(t) = \frac{\text{mean time waiting for service}}{\text{mean time being served}} = \frac{\text{time wasted}}{\text{useful time}}.$$

Using our distribution of $W(t)$ determined earlier, we calculate

$$R = \frac{\pi(N)}{N(1-\rho)^2}.$$

In order to reduce the customer loss ratio, management should allow for a substantial amount of idle time in a service facility at which demands for service occur randomly and the time required to render service is random, rather than attempting to attain a utilization factor as close to one as possible. For example, in a four-station service facility with utilization factor 90 percent, the customer loss ratio is 200 percent. If one more service station is added to the facility, the utilization factor is reduced to 72 percent and the customer loss ratio is less than 30 percent.

Some queues are set up on finite state spaces. Consider, for example, an N-server queue for telephone calls. If all N channels are busy any incoming calls are dropped. Here the state space is $\mathscr{L} = \{0, 1, \ldots, N\}$ and the birth and death rates are given by $\lambda_x = \lambda$ for $x < N$, $\lambda_N = 0$ and $\mu_x = x\mu$. We have

$$\lambda_y \gamma_y = y!\, \mu \left(\frac{\mu}{\lambda}\right)^{y-1}, \qquad y \in \mathscr{L};$$

and thus we arrive at *Erlang's Loss formula*

$$\pi(x) = c \frac{\left(\frac{\lambda}{\mu}\right)^x}{x!}, \qquad x \in \mathscr{L}$$

where

$$c^{-1} = \sum_{y=0}^{N} \frac{\left(\frac{\lambda}{\mu}\right)^y}{y!}.$$

Our next example is that of the *infinite server queue.* Based on our analysis of the N-server queue we find the birth and death rates to be $\lambda_x \equiv \lambda$ and $\mu_x = x\mu$, respectively, and the stationary distribution is Poisson with rate $\frac{\lambda}{\mu}$,

$$\pi(x) = e^{-\lambda/\mu} \frac{\left(\frac{\lambda}{\mu}\right)^x}{x!}.$$

(This queue is always positive recurrent.) For the infinite server queue we can actually calculate $P_{xy}(t)$ explicitly, but not by using the forward or backward equations.

To accomplish this, let $Y(t)$ denote the number of customers who arrive in the time interval $(0, t]$, and let $X_1(t)$ denote the number of these customers *still* in the process of being served at time t. Then we can write

$$X(t) = X_1(t) + X_2(t)$$

where $X_2(t)$ is the number of customers *present initially* still in the process of being served at time t. Under $\mathbb{P}_x$ the distribution of $X_2(t)$ is binomial with parameters x and $e^{-\mu t}$. To find the distribution of $X_1(t)$ we argue as follows. Conditioned on $Y(t)$ the arrival times in $(0, t]$ are independent $U(0, t)$. If a customer arrives at time $s \in (0, t]$ the probability he is still in the process of being served at time t is $e^{-\mu(t-s)}$. Thus if a customer arrives at a time chosen uniformly from $(0, t]$, the probability that he is still in the process of being served at time t is

$$p_t = \frac{1}{t}\int_0^t e^{-\mu(t-s)}\, ds = \frac{1 - e^{-\mu t}}{\mu t}.$$

Since $X_1(t)$ is a filtered version of $Y(t)$ with recording probability p_t we find that it is Poisson distributed with parameter $\lambda t p_t$. Finally, since $X_1(t)$ and $X_2(t)$ are independent we compute

$$\begin{aligned} P_{xy}(t) &= \sum_{k=0}^{\min(x,y)} \mathbb{P}_x(X_2(t) = k)\mathbb{P}(X_1(t) = y - k) \\ &= \sum_{k=0}^{\min(x,y)} \binom{x}{k} e^{-k\mu t}(1 - e^{-\mu t})^{x-k} \frac{\left[\frac{\lambda}{\mu}(1 - e^{-\mu t})\right]^{y-k}}{(y-k)!} e^{-(\lambda/\mu)(1-e^{-\mu t})} \end{aligned}$$

As $t \to \infty$ we check that $P_{xy}(t) \to \pi(y)$.

Our next example is that of a continuous-time *branching process.* Consider a collection of particles that act independently in giving rise to succeeding generations of particles. Suppose that each particle, from the time it appears, waits a random length of time having an exponential distribution with parameter μ and then splits into two identical particles with probability p, or else disappears altogether with probability $q = 1 - p$. Let $X(t)$ denote the number of particles present at time t. We calculate the parameters to be $q_x = x\mu$, $Q_{x\,x+1} = p$ and $Q_{x\,x-1} = q$. Thus the birth and death rates work out to be linear, $\lambda_x = x\mu p$ and $\mu_x = x\mu q$, respectively. State 0 is an absorbing state here.

We can modify the branching process by allowing for immigration. New particles immigrate into the system at random times, according to a Poisson process with parameter λ. The modified parameters work out to be $q_x = x\mu + \lambda$, $Q_{x\,x+1} = \dfrac{\lambda}{x\mu + \lambda} + \dfrac{x\mu}{x\mu + \lambda}p$ and $Q_{x\,x-1} = \dfrac{x\mu}{x\mu + \lambda}q$. Thus the modified birth and death rates are, respectively, $\lambda_x = \lambda + x\mu p$ and $\mu_x = x\mu q$. The terms γ_y are given by $\gamma_y = \prod_{x=1}^{y} \dfrac{x\mu q}{\lambda + x\mu p}$, and so the process is transient if $q < p$, null recurrent if $q = p$, and positive recurrent if $q > p$. The infinite server queue corresponds to branching with immigration where $p = 0$.

We consider next the *pure linear birth process*, with $\lambda_x = x\lambda$, $\mu_x \equiv 0$. One can show inductively from (8), or argue probabilistically, as we do later, that

$$P_{xy}(t) = \binom{y-1}{y-x} e^{-\lambda t x}(1 - e^{-\lambda t})^{y-x}, \qquad y \geq x.$$

That is, conditioned on $X(s)$, the random variable $X(s + t) - X(s)$ is negative binomial with parameters $r = X(s)$ and $p = e^{-\lambda t}$. This is explained as follows. The fact that $\lambda_x = x\lambda$ means that each of the x particles is independently growing at rate λ, like the Poisson process. Thus $X(s + t)$ is the sum of $X(s)$ i.i.d. random variables, each geometric with parameter $p = e^{-\lambda t}$. This process is called the *Yule process*. Using the geometric distribution we compute

$$\mathbf{E}[X(t)|X(0)] = X(0)e^{\lambda t}$$

$$\mathrm{Var}[X(t)|X(0)] = X(0)e^{\lambda t}(e^{\lambda t} - 1)$$

$$\mathbf{E}[e^{iuX(t)}|X(0)] = \left(\frac{e^{-\lambda t}}{e^{-iu} + e^{-\lambda t} - 1}\right)^{X(0)}$$

The Yule process corresponds to branching (without immigration), $p = 1$.

We focus next on the *linear birth and death process* (from Parzen [45]), with $\lambda_x = x\lambda$, $\mu_x = x\mu$. Here again, each of the x particles acts independently with birth rate λ and death rate μ. The state 0 is an absorbing state. Let Φ_x be the generating function

$$\Phi_x(z, t) = \mathbf{E}_x z^{X(t)}.$$

Fiz z and let u be the vector $u_y = z^y$, $y \in \mathscr{L}$. Then the evolution equation for Φ_x is

$$\frac{\partial \Phi_x}{\partial t} = \mathbf{E}_x(qu)_{X(t)},$$

where q is the generator. Now observe that

$$qu = [\mu - (\lambda + \mu)z + \lambda z^2]\frac{du}{dz} = (1 - z)(\mu - \lambda z)\frac{du}{dz}.$$

Thus we arrive at the partial differential equation for Φ_x,

$$\frac{\partial \Phi_x}{\partial t} = (1 - z)(\mu - \lambda z)\frac{\partial \Phi_x}{\partial z} \tag{12}$$

with initial condition

$$\Phi_x(z, 0) = z^x.$$

We set up the characteristics

$$\frac{dz}{dt} + (1 - z)(\mu - \lambda z) = 0$$

to solve (12). There are two cases:

Case I: $\lambda \neq \mu$

$$\Phi_x(z, t) = \left[1 + \frac{\mu - \lambda}{\lambda - \left(\lambda + \dfrac{\mu - \lambda}{1 - z}\right)e^{(\mu - \lambda)t}}\right]^x.$$

Case II: $\lambda = \mu$

$$\Phi_x(z, t) = \left[1 - \frac{1}{\dfrac{1}{1 - z} + \lambda t}\right]^x.$$

Since $P_{x0}(t)$ is the constant term $\Phi_x(0, t)$ we obtain the absorption probabilities

$$P_{x0}(t) = \left[1 - \frac{\lambda - \mu}{\lambda - \mu e^{(\mu - \lambda)t}}\right]^x \qquad \text{(Case I)}$$

$$P_{x0}(t) = \left(1 - \frac{1}{1 + \lambda t}\right)^x \qquad \text{(Case II).}$$

Exercises

1. (Hoel, Port and Stone [28]

$$q = \begin{bmatrix} -\lambda & \lambda & 0 \\ \mu & -\mu & 0 \\ 0 & \lambda & -\lambda \end{bmatrix}.$$

Find $P(t)$.

2. (Hoel, Port and Stone [28])

$$q = \begin{bmatrix} -\lambda & \lambda & 0 \\ \mu & -\mu - \eta & \eta \\ 0 & \lambda & -\lambda \end{bmatrix}$$

Find $P_{00}(t)$, $P_{01}(t)$, $P_{02}(t)$.

3. Let X_1, X_2 be independent exponentially distributed r.v.s with parameters λ_1, λ_2.

   (a) Find $P(X_1 < X_2)$.
   (b) Find the distribution of $\min(X_1, X_2)$.
   (c) Let $N = I_{\{X_1 < X_2\}}$. Prove that N and $\min(X_1, X_2)$ are independent.
   (d) Find the conditional pdf of $|X_1 - X_2|$ given N.
   (e) Show that $|X_1 - X_2|$ and $\min(X_1, X_2)$ are independent.

4. (Karlin and Taylor [31]) There are two transatlantic cables each of which can handle one telegraph message at a time. The time to breakdown for each has the same exponential distribution with parameter λ. The time to repair for each cable has the same exponential distribution with parameter μ. Given that initially both cables are working, find the probability that they will both be working at time t.

5. (Hoel, Port and Stone [28]) Suppose d particles are distributed into two boxes. A particle in box 0 remains in that box for a random length of time that is exponentially distributed with parameter λ before going to box 1. A particle in box 1 remains there for an amount of time that is exponentially distributed with parameter μ before going to box 0. The particles act independently of each other. Let $X(t)$ denote the number of particles in box 1 at time $t \geq 0$.

   (a) Find q.
   (b) Find $P_{xd}(t)$, $x = 0, 1, \ldots, d$.
   (c) Find $E_x X(t)$.
   (d) Find the stationary distribution.

6. (Hoel, Port and Stone [28]) Suppose d machines are subject to failures and repairs. The failure times are exponentially distributed with parameter μ, and the repair times are exponentially distributed with parameter λ. Let $X(t)$ denote the number of machines that are in satisfactory order at time t. Assume there is only one repairman. Find q and π.

7. $$q = \begin{bmatrix} -\lambda & \lambda \\ \mu & -\mu \end{bmatrix}.$$

   Let $N(t)$ be the number of times the system has changed states in time t. Find the pdf of $N(t)$.

8. (Karlin and Taylor [31]) A system is composed of N identical components; each independently operates a random length of time until failure. Suppose the failure time distribution is exponential with parameter λ. When a component fails, it undergoes repair. The repair time is random, with distribution function exponential with parameter μ. The system is said to be in state n at time t if there are exactly n components under repair at time t. Find q.

   Suppose initially all N components are operative. Find the mgf of the first time there are two inoperative components, T_2. In case $\lambda = \mu$ find the cdf.

9. (Karlin and Taylor [31]) Explosion:

$$q = \begin{bmatrix} -\lambda & \lambda & 0 & 0 & \cdots & \\ & -4\lambda & 4\lambda & 0 & \cdots & \\ & & -9\lambda & 9\lambda & \cdots & \\ & & & \ddots & \ddots & \\ & & & & -n^2\lambda & n^2\lambda \\ & & & & \ddots & \ddots \end{bmatrix}.$$

That is,

$$q_{xy} = \begin{cases} 0, & y < x \text{ or } y > x + 1 \\ -x^2\lambda, & y = x \\ x^2\lambda, & y = x + 1 \end{cases}$$

Find $P_1(X(t) = \infty)$.

10. What is $\lim_{t\uparrow\infty} P(t)$ in general, for a finite state space?

11. (Parzen [45]) For a Poisson point process in $\mathbb{R}^2$ find the distribution of D, the distance between an individual particle and its nearest neighbor. Find $\mathbf{E}D$. What about $\mathbb{R}^3$?

12. (Parzen [45]) A radioactive source is observed during four nonoverlapping time intervals of 6 seconds each. The number of particles emitted during each period is counted. If the particles emitted obey a Poisson law at a rate of 0.5 particles emitted per second, find the probability that

    (i) in each of the four time intervals three or more particles are counted;
    (ii) in at least one of the four time intervals three or more particles are counted.

13. (Parzen [45]) Let $\{X_1(t): t \geq 0\}$, $\{X_2(t): t \geq 0\}$ be independent Poisson processes with rates λ_1, λ_2, respectively. Set $X(t) = X_1(t) - X_2(t)$. Show that $\{X(t): t \geq 0\}$ has stationary independent increments, and find the pdf of $X(t)$. Compute $\lim_{t\uparrow\infty} \mathbb{P}(|X(t)| < c)$ for any $c > 0$.

14. (Parzen [45]) Consider a taxi station where taxis looking for customers and customers looking for taxis arrive in accord with Poisson processes, with mean rates per minute of 1 and 1.25. A taxi will wait no matter how many are in line. However, an arriving customer waits only if the number of customers already waiting for taxis is two or less.

    Find

    (i) the mean number of customers waiting for taxis;
    (ii) the mean number of taxis waiting for customers; and
    (iii) the mean number of customers who in the course of an hour do not join the waiting line because there were three customers already waiting.

15. (Hoel, Port and Stone [28]) Let $\{X(t): t \geq 0\}$ be a Poisson process with rate λ. Let T be Gamma (α, ν), independent of X. Find the distribution of $X(T)$.

16. (Hoel, Port and Stone [28]) Let $\{X(t): t \geq 0\}$ be a Poisson process with rate λ. Find the conditional distribution of τ_1, given $X(t)$.

17. Let $\{X(t): t \geq 0\}$ be a Poisson process with rate λ. Find $\mathrm{Cov}(X(s), X(t))$.

18. (Karlin and Taylor [30]) Let $\{X(t): t \geq 0\}$, $\{Y(t): t \geq 0\}$ be Poisson processes with rates λ, μ. Find the distribution of the number, N, of jumps of the Y process, which occur between successive jumps of the X process.

19. (Karlin and Taylor [31]) Let $(X(t), Y(t))$ describe a stochastic process in two dimensions, where X and Y are independent Poisson processes with rates λ, μ. Given that the process is in state (x_0, y_0) what is the probability that it will intersect the line $x + y = z$ at the point (x, y)? (Assume $z > x_0 + y_0$.)

20. (Karlin and Taylor [31]) Let $\{X(t): t \geq 0\}$ be a Poisson process with rate λ. Find the pdf of τ_k given $X(t) = n$. (Take $k < n$.)

Section VII
Ergodic Theory with an Application to Fractals

Ergodic Theorems

We have already seen examples of limit theorems in Sections V and VI that assert the convergence of temporal averages to spatial averages. Thus if x is a positive recurrent state of an aperiodic irreducible Markov chain, then $\lim_{n\to\infty} N_n(x)/n = \pi(x)$. That is, the *temporal* average fraction of time the chain spends at state x converges to the *spatial* average $\pi(x)$. Similarly, for Markov jump processes $\frac{1}{t}\int_0^t I_{\{x\}}(X(s))\,ds \to \pi(x)$ as $t \to \infty$, when x is positive recurrent.

Ergodic theory is a general setting in which this same type of phenomenon holds; that is, where

$$\lim_{n\to\infty} \frac{1}{n}\sum_{k=1}^{n} f(X_k) = \int f(x)\,dF(x)$$

for a discrete time stochastic process $\{X_n\}_{n\geq 1}$, or

$$\lim_{t\to\infty} \frac{1}{t}\int_0^t f(X(s))\,ds = \int f(x)\,dF(x)$$

for a continuous time stochastic process $\{X(t): t \geq 0\}$. Here F is the d.f. for some stationary distribution. We shall confine ourselves to the discrete time case and concentrate on stationary sequences $\{X_n\}_{n\geq 1}$. The sequence $\{X_n\}_{n\geq 1}$ of random variables is *stationary* if for every $k, n \in \mathbb{N}$ $(X_1, \ldots, X_n)$ and $(X_{k+1}, \ldots, X_{k+n})$ have the same joint distribution. In particular, of course, the individual random variables X_n are all identically distributed. They need not be independent, however, although an i.i.d. sequence is always stationary. In Section III, we defined the notion of a tail event for a sequence $\{X_n\}_{n\geq 1}$, which is an event determined by $X_N, X_{N+1}, \ldots$ alone, for any N. A stronger notion is that of an *invariant event*, which is an event that has the *same* form

$$(X_N, X_{N+1}, \ldots) \in B$$

for any N. Thus, a tail event like

$$\left\{\lim_{n\to\infty} \frac{1}{n} \sum_{k=1}^{n} X_k = \mu\right\}$$

is invariant, whereas a tail event like

$$\left\{\limsup_{n\to\infty} \frac{X_{3n+2} - 5}{2^n} \leq 10\right\}$$

is not invariant (since it depends on a particular subsequence of X's). The stationary sequence $\{X_n\}_{n\geq 1}$ is *ergodic* if every invariant event has probability zero or one. In particular, an i.i.d. sequence is always stationary ergodic by Kolmogorov's Zero-One Law.

One version of the Ergodic Theorem goes as follows.

Pointwise Ergodic Theorem. *Let* $\{X_n\}_{n\geq 1}$ *be a stationary ergodic sequence of random variables with* $\mathbb{E}|X_n| < \infty$. *Then*

$$\lim_{n\to\infty} \frac{1}{n} \sum_{k=1}^{n} X_k = \mathbb{E}X, \qquad \text{a.s.}$$

The proof of this result relies on the following.

Maximal Ergodic Theorem. *Let* $\{X_n\}_{n\geq 1}$ *be a stationary sequence of random variables with* $\mathbb{E}|X_n| < \infty$. *Define*

$$S_n = \sum_{k=1}^{n} X_k, \qquad M_n = \max(0, S_1, S_2, \ldots, S_n).$$

Then for any $n \in \mathbb{N}$

$$\mathbb{E}I_{\{M_n > 0\}} X_1 \geq 0.$$

PROOF. Define

$$S'_n = \sum_{k=2}^{n+1} X_k, \qquad M'_n = \max(0, S'_1, S'_2, \ldots, S'_n).$$

Since $(X_1, \ldots, X_n)$ and $(X_2, \ldots, X_{n+1})$ have the same joint distribution, it follows that M_n and M'_n have the same distribution. In particular,

$$\mathbb{E}M_n = \mathbb{E}M'_n.$$

Since

$$X_1 + M'_n \geq \max(S_1, S_2, \ldots, S_n)$$

we can estimate

$$\begin{aligned}\mathbb{E}I_{\{M_n>0\}}X_1 &\geq \mathbb{E}I_{\{M_n>0\}}[\max(S_1, S_2, \ldots, S_n) - M_n'] \\ &= \mathbb{E}I_{\{M_n>0\}}(M_n - M_n') = \mathbb{E}M_n - \mathbb{E}I_{\{M_n>0\}}M_n' \\ &\geq \mathbb{E}M_n - \mathbb{E}M_n' = 0.\end{aligned}$$

□

Proof of Ergodic Theorem. By replacing X_n with $X_n - \mathbb{E}X_n$, we can assume without loss of generality that $\mathbb{E}X_n = 0$. For any $\varepsilon > 0$ let D be the invariant event

$$D = \left\{\limsup_n \frac{S_n}{n} > \varepsilon\right\},$$

where $S_n = \sum_{k=1}^n X_k$. Define

$$X_n' = I_D(X_n - \varepsilon).$$

Since D is invariant, the sequence $\{X_n'\}_{n\geq 1}$ is also stationary. Thus, if we define

$$S_n' = \sum_{k=1}^n X_k', \qquad M_n' = \max(0, S_1', S_2', \ldots, S_n'),$$

then the Maximal Ergodic Theorem gives

$$\mathbb{E}I_{\{M_n'>0\}}X_1' \geq 0.$$

It follows from the Dominated Convergence Theorem that

$$\mathbb{E}I_D X_1' = \lim_{n\to\infty} \mathbb{E}I_{\{M_n'>0\}}X_1' \geq 0,$$

since

$$\left\{\sup_{n\geq 1} S_n' > 0\right\} = \left\{\sup_{n\geq 1} \frac{S_n}{n} > \varepsilon\right\} \cap D = D$$

On the other hand,

$$\mathbb{E}I_D X_1' = \mathbb{E}I_D X_1 - \varepsilon\mathbb{P}(D).$$

Since $\mathbb{P}(D)$ must equal zero or one, and since $\mathbb{E}X_1 = 0$, we conclude that in fact $\mathbb{P}(D) = 0$. Since $\varepsilon > 0$ is arbitrary, we conclude that $\lim_n \sup S_n/n \leq 0$ a.s. Applying the same argument to the sequence $\{-X_n\}_{n\geq 1}$ leads us to conclude that in fact

$$\lim_n S_n/n = 0 \qquad \text{a.s.}$$

□

The Ergodic Theorem is due to Birkhoff. There is a Hilbert space version due to von Neumann, which we will describe next. Let $T\colon H \to H$ be a bounded linear operator on a Hilbert space H. Denote by $N(T)$ and

$R(T)$ the null space and range of T, respectively. We say that T is a *contraction* if $\|T\| \leq 1$.

Mean Ergodic Theorem. *Let* T *be a contraction on a Hilbert space* H *and let* $S = N(I - T)$. *Then* $\frac{1}{n}\sum_{k=1}^{n} T^k x$ *converges in norm to* $P_S x$ *as* $n \to \infty$ *for all* $x \in H$.

Proof. Observe first that

$$N(I - T) = N(I - T^*) \tag{1}$$

where T^* is the adjoint of T, since T is contractive. Indeed, if $\langle y, Ty \rangle$ is real, then $\langle y, Ty \rangle = \langle Ty, y \rangle$; and thus

$$\begin{aligned} \|Ty - y\|^2 &= \|Ty\|^2 + \|y\|^2 - 2\langle Ty, y \rangle \\ &\leq 2\|y\|^2 - 2\langle Ty, y \rangle. \end{aligned}$$

From this, it follows that $Ty = y$ if and only if $\langle y, Ty \rangle = \|y\|^2$. But since T^* is also contractive, we get that same condition for $T^*y = y$.

For any bounded linear operator $A: H \to H$ there holds

$$N(A)^{\perp} = \overline{R(A^*)}.$$

(Just think about the inclusions $R(A^*) \subseteq N(A)^{\perp}$ and $R(A^*)^{\perp} \subset N(A)$.) In particular, using (1),

$$N(I - T)^{\perp} = N(I - T^*)^{\perp} = \overline{R(I - T)}.$$

Thus any $x \in H$ can be decomposed as

$$x = y + w \tag{2}$$

where $Ty = y$ and $w \in \overline{R(I - T)}$.

If $Ty = y$ then

$$\frac{1}{n}\sum_{k=1}^{n} T^k y = y.$$

If $w = z - Tz$ for some $z \in H$, then

$$\sum_{k=1}^{n} T^k w = Tz - T^{n+1}z;$$

since $\|T\| \leq 1$ it is clear that

$$\frac{1}{n}\sum_{k=1}^{n} T^k w \to 0 \qquad \text{as } n \to \infty.$$

By approximation, again since $\|T\| \leq 1$, this extends to $w \in \overline{R(I - T)}$. Thus by (2)

$$\lim_{n\to\infty} \frac{1}{n}\sum_{k=1}^{n} T^k x = y = P_S x. \qquad \square$$

To see the connection between the Birkhoff and von Neumann Ergodic Theorems, take $H = L^2(\mathbb{P})$, and let T be the *shift operator*

$$TX(\omega) = X(\tau\omega)$$

where τ is a *measure-preserving transformation*. That is, X and $X \circ \tau$ have the same distribution for any $X \in H$. The sequence $\{X_n\}_{n \geq 1}$ in the Pointwise Ergodic Theorem corresponds to $\{X \circ \tau^{n-1}\}_{n \geq 1}$, and $S = N(I - T)$ is the subspace of invariant random variables; i.e., random variables Y satisfying $Y(\omega) = Y(\tau\omega)$ a.s. Then $P_S X$ is the closest invariant random variable to X; i.e., the conditional expectation of X given S. When τ is ergodic the invariant random variables are constant a.s., in which case $P_S X$ is just the constant $\mathbb{E}X$.

In fact, the Pointwise Ergodic Theorem carries over to the case where the sequence $\{X_n\}$ is only assumed to be stationary (not necessarily ergodic). The conclusion generalizes to

$$\lim_{n \to \infty} \frac{1}{n} \sum_{k=1}^{n} X_k = \mathbb{E}(X_1 | S) \tag{3}$$

where $\mathbb{E}(X_1 | S) = P_S X_1$ is the closest random variable in S to X_1. This conditional expectation satisfies the important property

$$\mathbb{E} Y X_1 = \mathbb{E}[Y \mathbb{E}(X_1 | S)] \tag{4}$$

for all $Y \in S$. (Compare (10) in Section II). In particular, if D is an invariant event, then $Y = I_D \in S$, and

$$\mathbb{E} I_D X_1 = \mathbb{E}[I_D \mathbb{E}(X_1 | S)].$$

The proof of (3) follows the same lines as the proof of the Pointwise Ergodic Theorem.

Subadditive Ergodic Theorem

A far-reaching generalization of the Ergodic Theorem is *Kingman's Subadditive Ergodic Theorem.* Here we deal with two-parameter stochastic processes $\{X_{mn}: 0 \leq m < n\}$. Such a process is *subadditive* if

(S_1) $\qquad X_{ln} \leq X_{lm} + X_{mn}, \qquad l < m < n;$

(S_2) $\{X_{m+1,n+1}\}$ *has the same distribution as* $\{X_{mn}\}$;

(S_3) $\gamma_n = \mathbb{E}X_{0n}$ *exists and satisfies* $\gamma_n \geq -An$, $n \geq 1$, *for some constant* A.

It follows from (S_2) that $\mathbb{E}X_{mn} = \gamma_{n-m}$, and taking expectation as in (S_1) leads to

$$\gamma_{m+n} \leq \gamma_m + \gamma_n.$$

This implies that

$$\lim_{n\to\infty} \frac{\gamma_n}{n} = \gamma$$

where $\gamma = \inf_n \frac{\gamma_n}{n}$ is finite, on account of (S_3). Denote $\gamma = \gamma(X)$; it is called the *time constant* of X.

The process $\{X_{mn}\}$ is *additive* if (S_1) is strengthened to equality:

$$X_{ln} = X_{lm} + X_{mn}.$$

In this case, the process $\{Y_n\}_{n\geq 1}$ defined by $Y_n = X_{n-1,n}$ is stationary, and

$$X_{0n} = \sum_{k=1}^{n} Y_k. \tag{5}$$

Subadditive Ergodic Theorem. *If $\{X_{mn}\}$ is a subadditive process with time constant γ, then the finite limit*

$$\xi = \lim_{n\to\infty} \frac{X_{0n}}{n}$$

exists a.s., and $\mathbb{E}\xi = \gamma$.

Observe that if $\{X_{mn}\}$ is additive, then this theorem follows from the Pointwise Ergodic Theorem, on account of (5). Moreover, $\gamma = \mathbb{E}Y$, in this case.

The proof of the Subadditive Ergodic Theorem relies on the following theorem, which we state here without proof.

Decomposition Theorem. *A subadditive process $\{X_{mn}\}$ admits a decomposition*

$$X_{mn} = Y_{mn} + Z_{mn}$$

where $\{Y_{mn}\}$ is additive, and $Z_{mn} \geq 0$ with $\gamma(Z) = 0$.

Proof of Subadditive Ergodic Theorem. As mentioned earlier, the Subadditive Ergodic Theorem holds for the additive process $\{Y_{mn}\}$. So, on account of the Decomposition Theorem, we may assume that $X_{mn} \geq 0$ and $\gamma = 0$. Set

$$\xi = \limsup_{n\to\infty} \frac{X_{0n}}{n}.$$

Then it suffices to show that $\xi = 0$ a.s.; to do this it suffices to show that $\mathbb{E}\xi = 0$ (since $\xi \geq 0$).

Fix r. Repeated application of (S_1) gives

$$X_{0n} \leq \sum_{k=1}^{N} X_{(k-1)r,kr} + W_N$$

where $N = \left[\frac{n}{r}\right]$ and

$$W_N = \sum_{k=1}^{r-1} |X_{Nr,Nr+k}|.$$

Since the one-parameter process $\{X_{(k-1)r,kr}: k \geq 1\}$ is stationary, it follows from the Ergodic Theorem that

$$X_r^* = \lim_{N\to\infty} \frac{1}{N} \sum_{k=1}^{N} X_{(k-1)r,kr}$$

exists a.s., and $\mathbb{E}X_r^* = \gamma_r$. Moreover $\{W_N\}_{N\geq 1}$ is stationary with finite mean, and so

$$\lim_{N\to\infty} \frac{W_N}{N} = 0 \qquad \text{a.s.}$$

Therefore

$$\begin{aligned} \xi &\leq \limsup_{n\to\infty} \frac{X_{on}}{rN} \\ &\leq \limsup_{n\to\infty} \frac{1}{rN}\left[\sum_{k=1}^{N} X_{(k-1)r,kr} + W_N\right] = X_r^*/r. \end{aligned}$$

In particular, $\mathbb{E}\xi \leq \gamma_r/r$. Since r was arbitrary and $\gamma = 0$, we conclude that $\mathbb{E}\xi = 0$. □

For some purposes (S_3) can be weakened to

(S_3') $$\mathbb{E}X_{01}^{+} < \infty$$

where x^+ denotes $\max(0, x)$.

Corollary I. *If* $\{X_{mn}\}$ *satisfies* (S_1), (S_2), and (S_3') *but not* (S_3) *then the limit*

$$\xi = \lim_{n\to\infty} \frac{X_{on}}{n}$$

exist a.s., $-\infty \leq \xi < \infty$ *and* $\mathbb{E}\xi = -\infty$.

PROOF. Fix N and set

$$X_{mn}^{(N)} = \max(X_{mn}, -N(n-m))$$

Observe that $\{X_{mn}^{(N)}\}$ is a subadditive process obeying (S_3). Thus, by the

Subadditive Ergodic Theorem

$$\xi^{(N)} = \lim_{n\to\infty} \frac{X_{0n}^{(N)}}{n}$$

exists a.s. Since $\dfrac{X_{0n}^{(N)}}{n} = \max\left(\dfrac{X_{0n}}{n}, -N\right)$, it follows that ξ exists, and

$$\xi^{(N)} = \max(\xi, -N).$$

Hence $\mathbb{E}\xi \leq \mathbb{E}\xi^{(N)} \leq \dfrac{1}{n}\mathbb{E}X_{0n}^{(N)}$. Letting $N \to \infty$ and then $n \to \infty$ leads to $\mathbb{E}\xi \leq -\infty$. □

When dealing with two-parameter processes, an *invariant event* is defined as an event invariant under the shift $\{X_{mn}\} \to \{X_{m+1,n+1}\}$. That is, it is an event that has the same form

$$(X_{m+N,n+N}: 0 \leq m < n) \in B$$

for any N. The subadditive process $\{X_{mn}\}$ is *ergodic* if the probability of any invariant event is either zero or one. By tracking through the preceding proof of the Subadditive Ergodic Theorem, it can be shown that *if* $\{X_{mn}\}$ *is ergodic then the limit* ξ *is constant a.s.*; namely,

$$\xi \equiv \gamma \qquad \text{a.s.}$$

Products of Random Matrices

One of the main applications of the Subadditive Ergodic Theorem is to *products of random matrices.*

Corollary II. *Let* $A_1, A_2, \ldots$ *be a stationary ergodic sequence of random* $d \times d$ *matrices, with*

$$\mathbb{E}\log^+ \|A_1\| < \infty.$$

Then

$$\lambda = \lim_{n\to\infty} \frac{1}{n}\log \|A_n \ldots A_1\| \tag{6}$$

exists a.s. and is constant, $-\infty \leq \lambda < \infty$.

Proof. For $m \leq n$ denote ${}^nA^m = A_n \ldots A_m$. The process

$$\{X_{mn} = \log \|{}^nA^{m+1}\|: 0 \leq m < n\}$$

is subadditive and ergodic; i.e., it obeys (S_1), (S_2), and (S_3'). Thus by our preceding corollary to the Subadditive Ergodic Theorem, we conclude that

$$\xi = \lim_{n\to\infty} \frac{X_{0n}}{n}$$

exists a.s., $-\infty \le \xi < \infty$. Since $\{X_{mn}\}$ is ergodic we also conclude that $\xi = \gamma$ a.s. □

Corollary III. *Let* $A_1, A_2, \ldots$ *be a stationary sequence of* $d \times d$ *matrices with strictly positive entries, and assume that* $\mathbb{E}\log(A_1)_{ij}$ *exists and is finite,* $1 \le i, j \le d$. *Then the finite limit*

$$\xi = \lim_{n\to\infty} \frac{1}{n} \log({}^{n}A^{1})_{ij}$$

exists a.s.

Proof. Set

$$Y_{mn} = ({}^{n}A^{m+1})_{1,1} \qquad 0 \le m < n.$$

This time $Y_{ln} \ge Y_{lm} Y_{mn}$, and so the process

$$\{X_{mn} = -\log Y_{mn} \colon 0 \le m < n\}$$

satisfies (S_1). Since (A_n) is stationary, (S_2) holds. To see that (S_3) holds we have $\frac{\gamma_n}{n} \ge -\mathbb{E}\log\|A_1\|$, where we use the l_1-norm $\|A\| = \max_i \sum_j |A_{ij}|$ to estimate

$$\mathbb{E}\log\|A_1\| = \mathbb{E}\max_i \log \sum_j (A_1)_{ij}$$

$$\le \mathbb{E}\max_{i,j} \log d(A_1)_{ij} \le \mathbb{E}\sum_{i,j} \log^+(A_1)_{ij} + \log d < \infty.$$

Applying the Subadditive Ergodic Theorem to $\{X_{mn}\}$ establishes the Corollary for $i = j = 1$.

To handle the other values of i, j we use the inequalities

$$({}^{n+1}A^{2})_{ij} \ge (A_{n+1})_{i1} Y_{2,n} (A_2)_{1j}$$

$$Y_{0,n+2} \ge (A_{n+2})_{1j} ({}^{n+1}A^{2})_{ij} (A_1)_{j1}.$$

Since $\mathbb{E}\log(A_n)_{i1}$ is finite, $\frac{1}{n}\log(A_n)_{i1} \to 0$ a.s. Thus by taking logs in these inequalities, dividing by n, and letting $n \to \infty$, we find that $\frac{1}{n}\log({}^{n+1}A^{2})_{ij} \to \xi$ a.s. By stationarity the same holds for ${}^{n}A^{1}$. □

To see that strict positivity is essential in this corollary, consider an i.i.d. sequence (A_n) of 2×2 matrices, with

$$A = \begin{cases} \begin{bmatrix} 0 & 1 \\ 1 & 0 \end{bmatrix}, & \text{w.p. } 1/2 \\ \begin{bmatrix} 2 & 0 \\ 0 & 1 \end{bmatrix}, & \text{w.p. } 1/2. \end{cases}$$

Then

$${}^nA^1 = \begin{bmatrix} 2^{k_1} & 0 \\ 0 & 2^{k_2} \end{bmatrix} \quad \text{or} \quad \begin{bmatrix} 0 & 2^{k_2} \\ 2^{k_1} & 0 \end{bmatrix}$$

where $\dfrac{k_1 + k_2}{n} \to 1/2$.

The constant λ in (6) can be thought of as the *exponential growth rate* for the products ${}^nA^1$. When (A_n) is i.i.d. then $\rho = e^\lambda$ is sometimes called the *spectral radius* for the (common) distribution of the A_ns. This is because in the nonrandom case where $A_1 = A_2 = \cdots = A$, then $\lim_{n\to\infty} \|A^n\|^{1/n} = \rho(A)$.

Although in the i.i.d. case λ is a function of the distribution of the A_ns, in general it is very difficult to compute. Sometimes, as in stability theory, one is only concerned with whether or not $\lambda < 0$. Observe that if $\lambda < 0$ then

$$\lim_{n\to\infty} A_n \dots A_1 = 0 \qquad \text{a.s.}$$

We do know that

$$\lambda = \lim_{n\to\infty} \frac{1}{n} \mathbb{E} \log \|{}^nA^1\| = \inf_n \frac{1}{n} \mathbb{E} \log \|{}^nA^1\|.$$

Thus the condition that $\lambda < 0$ is equivalent to

$$\mathbb{E} \log \|{}^nA^1\| < 0 \qquad \textit{for some } n \in \mathbb{N}.$$

Sometimes one has a whole family of distributions on $d \times d$ matrices, one for each d, and is interested in how λ depends on the dimension d asymptotically. In special cases explicit formulas for λ are known.

Example

Let $\{A_n\}_{n\geq 1}$ be i.i.d. with entries $((A_n)_{ij}: 1 \leq i, j \leq d)$ that are i.i.d. $N(0, 1)$. Set $\psi(t) = \Gamma'(t)/\Gamma(t)$. Then for any $\mathbf{x} \neq \mathbf{0}$

$$\lim_{n\to\infty} \frac{1}{n} \log \|{}^nA^1\mathbf{x}\| = \lambda_d$$

$$\frac{1}{\sqrt{n}}(\log \|{}^nA^1\mathbf{x}\| - n\lambda_d) \xrightarrow{D} N(0, \sigma_d^2)$$

where

$$\lambda_d = \frac{\psi(d/2) + \log 2}{2}, \qquad \sigma_d^2 = \frac{1}{4}\psi'\left(\frac{d}{2}\right).$$

To see this, observe that if $U \in \mathbb{R}^d$ is a random variable independent of A_1, $\|U\| = 1$ (l_2-norm), then $A_1 U$ is also independent of U. This is because each entry of $A_1 \mathbf{u}$ is standard normal, for any $\|\mathbf{u}\| = 1$. Thus the sequence

$$\|{}^nA^1\mathbf{x}\|/\|{}^{n-1}A^1\mathbf{x}\|: n \geq 1$$

is i.i.d., and so

$$\log \|{}^nA^1\mathbf{x}\| = \sum_{k=1}^{n} X_k + \log \|\mathbf{x}\|$$

where $\{X_k\}_{k\geq 1}$ is an i.i.d. sequence, each X_k distributed like $\frac{1}{2}\log \chi^2(d)$. Now use LLN and CLT.

Observe that $\lim_{d\to\infty} e^{\lambda_d}/\sqrt{d} = 1$, so that *the spectral radius grows like* $\sqrt{d}$. More generally, one can replace the $N(0, 1)$ distribution of the entries $(A_n)_{ij}$ with the distribution of W, where

$$\varphi_W(u) = \exp(-|u|^\alpha/\alpha)$$

and work with the l_α-norm. (For example, $\alpha = 1$ gives rise to the Cauchy distribution.) The same results hold, only now

$$\lambda_{d,\alpha} = \frac{1}{\alpha}\,\mathbb{E}\log S; \qquad \sigma_{d,\alpha}^2 = \frac{1}{\alpha^2}\operatorname{Var}\log S$$

where $S \sim |W_1|^\alpha + \cdots + |W_d|^\alpha$. It can be shown that

$$\lim_{d\to\infty} \frac{e^{\lambda_{d,\alpha}}}{(d\log d)^{1/\alpha}} = \left[\frac{2\Gamma(\alpha)\sin(\alpha\pi/2)}{\alpha\pi}\right]^{1/\alpha}.$$

Oseledec's Theorem

It turns out that (6) is the beginning of a full spectral decomposition for products of random matrices. The main result in this direction is *Oseledec's Theorem*, which we state and then prove in a sequence of steps.

Oseledec's Theorem. *Let* $\mathrm{A}_1, \mathrm{A}_2, \ldots$ *be a stationary ergodic sequence of* $\mathrm{d} \times \mathrm{d}$ *matrices, with*

$$\mathbb{E}\log^+\|A_1\| < \infty.$$

Then

(i) $\lim_{n\to\infty} [({}^{\mathrm{n}}\mathrm{A}^1)^*({}^{\mathrm{n}}\mathrm{A}^1)]^{1/2\mathrm{n}} = \Lambda$ *exists a.s., where* ${}^{\mathrm{n}}\mathrm{A}^1 = \mathrm{A}_{\mathrm{n}}\ldots\mathrm{A}_1$

(ii) *Let Λ have the (random) spectral decomposition*

$$\Lambda = \sum_{k=1}^{r} e^{\lambda'_k} P_{E_k}, \qquad \lambda'_1 < \cdots < \lambda'_r.$$

Then r, λ'_k, and $\dim(E_k)$ are constant a.s. Moreover if

$$V_k = \bigoplus_{i=1}^{k} E_i$$

then

$$\lim_{n\to\infty} \frac{1}{n} \log \|{}^nA^1\mathbf{v}\| = \lambda'_k \qquad a.s., \mathbf{v} \in V_k \setminus V_{k-1}.$$

PROOF. The proof of this result follows a number of steps.

Step I

$$\lim_{n\to\infty} \frac{1}{n} \log \mu_i(n) = \lambda_i \tag{7}$$

exists a.s., $1 \leq i \leq d$, and is constant a.s., where $\mu_1^2(n) \leq \cdots \leq \mu_d^2(n)$ are the eigenvalues of $({}^nA^1)^*({}^nA^1)$, and $-\infty \leq \lambda_i < \infty$.

PROOF. Recall for $1 \leq p \leq d$ the p-fold exterior power $\bigwedge^p \mathbb{R}^d$, which is the set of all formal expressions

$$\Sigma c_i(\mathbf{u}_{i_1} \wedge \cdots \wedge \mathbf{u}_{i_p})$$

where $c_i \in \mathbb{R}$; $\mathbf{u}_{i_1}, \ldots, \mathbf{u}_{i_p} \in \mathbb{R}^d$ with the conventions

(a) $\mathbf{u}_1 \wedge \cdots \wedge (\mathbf{u}_j + \mathbf{u}'_j) \wedge \cdots \wedge \mathbf{u}_p$
$= \mathbf{u}_1 \wedge \cdots \wedge \mathbf{u}_j \wedge \cdots \wedge \mathbf{u}_p + \mathbf{u}_1 \wedge \cdots \wedge \mathbf{u}'_j \wedge \cdots \wedge \mathbf{u}_p$.
(b) $c(\mathbf{u}_1 \wedge \cdots \wedge \mathbf{u}_p) = \mathbf{u}_1 \wedge \cdots \wedge c\mathbf{u}_j \wedge \cdots \wedge \mathbf{u}_p$.
(c) $\mathbf{u}_{\pi_1} \wedge \cdots \wedge \mathbf{u}_{\pi_p} = \text{sign}(\pi)\mathbf{u}_1 \wedge \cdots \wedge \mathbf{u}_p$.

This is an inner product space with

$$\langle \mathbf{u}_1 \wedge \cdots \wedge \mathbf{u}_p, \mathbf{v}_1 \wedge \cdots \wedge \mathbf{v}_p \rangle = \det(\langle \mathbf{u}_i, \mathbf{v}_j \rangle).$$

The p-fold exterior power $\bigwedge^p A$ of a $d \times d$ matrix A is given by

$$\bigwedge^p A: \mathbf{u}_1 \wedge \cdots \wedge \mathbf{u}_p \mapsto A\mathbf{u}_1 \wedge \cdots \wedge A\mathbf{u}_p.$$

Observe that

$$\left(\bigwedge^p A\right)\left(\bigwedge^p B\right) = \bigwedge^p (AB)$$

$$\left(\bigwedge^p A\right)^* = \bigwedge^p A^*.$$

Furthermore, the eigenvalues of $\bigwedge^p A$ are $\{\lambda_{i_1} \dots \lambda_{i_p} : 1 \le i_1 < \cdots < i_p \le d\}$, where $\lambda_1, \dots, \lambda_d$ are the eigenvalues of A.

Since $\mu_d(n) = \|{}^nA^1\|$ we establish (7) for $i = d$ by applying (6) directly. If we apply (6) to the stationary ergodic sequence $\left\{\bigwedge^p A_n\right\}_{n \ge 1}$ as we did to $\{A_n\}_{n \ge 1}$, we deduce that

$$\lim_{n \to \infty} \frac{1}{n} \log \left\| {}^n\!\left(\bigwedge^p A\right)^1 \right\| = \lim_{n \to \infty} \frac{1}{n} \left\| \bigwedge^p ({}^nA^1) \right\|$$

exists a.s. and is constant a.s. Thus

$$\lim_{n \to \infty} \frac{1}{n} \log[\mu_d(n) \dots \mu_{d-p+1}(n)]$$

exists a.s. for each $1 \le p \le d$, and we arrive at (7) for $i = d - p + 1$, having already proved it for $d, \dots, d - p + 2$. $\square$

Step II

$$\lim_{n \to \infty} \frac{1}{n} \log^+ \|A_n\| = 0 \qquad a.s. \tag{8}$$

PROOF. This follows from the Ergodic Theorem (or the Borel-Cantelli Lemma). $\square$

The polar decomposition of matrices shows that any matrix A can be written in the form

$$A = K'(A^*A)^{1/2}$$

with K' orthogonal. Since any symmetric nonnegative definite matrix B can be written in the form

$$B = K_1 D K$$

where D is a diagonal matrix with nonnegative entries and K_1 and K are orthogonal, we see that A can in fact be written in the form

$$A = LDK$$

with D diagonal (nonnegative entries) and K and L orthogonal. Furthermore, the diagonal entries of D are the eigenvalues of $(A^*A)^{1/2}$. Applying this now to ${}^nA^1$ we can write

$${}^nA^1 = L_n D_n K_n$$

where $D_n = \operatorname{diag}(\mu_1(n), \dots, \mu_d(n))$, and thus $({}^nA^1)^*({}^nA^1) = K_n^* D_n^2 K_n$.

Step III

For any $\varepsilon > 0$ *there exists* $N = N(\varepsilon)$ *such that whenever* $n \geq N$, $k \geq 0$

$$|(K_n K^*_{n+k})_{ij}| \leq \begin{cases} \exp(-n|\lambda_i - \lambda_j| + n\varepsilon), \\ \lambda_i, \lambda_j > -\infty \quad or \quad \lambda_i = \lambda_j = -\infty, \\ e^{-n/\varepsilon}, \quad otherwise. \end{cases}$$

In particular if K *and* $\hat{K}$ *are accumulation points of* $\{K_n\}$ *then* $(\hat{K}K^*)_{ij} = 0$ *if* $\lambda_i \neq \lambda_j$.

The proof of this step itself takes two substeps.

Step III(a)

For any $\varepsilon > 0$, $k \geq 0$ *there exists* $N = N(\varepsilon, k)$ *such that whenever* $n \geq N$

$$|(K_n K^*_{n+k})_{ij}| \leq \begin{cases} \exp(-n|\lambda_i - \lambda_j| + n\varepsilon), \\ \lambda_i, \lambda_j > -\infty \quad or \quad \lambda_i = \lambda_j = -\infty, \\ e^{-n/\varepsilon}, \quad otherwise. \end{cases}$$

Proof. Assume first that $\lambda_i \leq \lambda_j$. Since

$$D_{n+k} K_{n+k} K^*_n = L^*_{n+k} A_{n+k} \ldots A_{n+1} L_n D_n$$

it follows from Step II that

$$\mu_j(n+k)|(K_n K^*_{n+k})_{ij}| \leq \mu_i(n) e^{nk\varepsilon}$$

for n sufficiently large. Thus

$$|(K_n K^*_{n+k})_{ij}| \leq \exp[n(\lambda_i - \lambda_j) + n\varepsilon]. \tag{9}$$

Observe that if $\lambda_i > \lambda_j$ then (9) holds trivially.

For $\lambda_i > \lambda_j$ we argue that $(K_n K^*_{n+k})_{ij}$ equals its (i, j)-cofactor. Thus by applying (9) to each entry in this cofactor we see that

$$\begin{aligned} |(K_n K^*_{n+k})_{ij}| &\leq (d-1)! \exp\left[(d-1)n\varepsilon + n\sum_{l \neq i} \lambda_l - n\sum_{l \neq j} \lambda_j\right] \\ &= (d-1)! \exp[(d-1)n\varepsilon + n(\lambda_j - \lambda_i)]. \end{aligned}$$

□

Step III(b)

Let $r_i = \#\{j: \lambda_j \geq \lambda_i\}$. *Then for any* $\varepsilon > 0$ *there exists* $N = N(\varepsilon)$ *such that whenever* $n \geq N$, $k \geq 0$

$$|(K_n K^*_{n+k})_{ij}| \leq \begin{cases} \exp(-n|\lambda_i - \lambda_j| + 2|r_i - r_j|n\varepsilon), \\ \lambda_i, \lambda_j > -\infty \quad or \quad \lambda_i = \lambda_j = -\infty, \\ e^{-n/\varepsilon}, \quad otherwise. \end{cases}$$

PROOF. The proof of this estimate proceeds by induction on k. We will choose N so that the estimate in Step III(a) holds for $k = M$ and then show that if the estimate here holds for $n \geq N$ and some fixed $k = k_*$, then in fact it also holds for $n \geq N$ and $k = k_* + M$. To carry this out, M will have to be chosen appropriately. Precisely, we will show that if $\varepsilon < \dfrac{\rho}{4d}$ where

$$\rho = \min(|\lambda_i - \lambda_j| : \lambda_i \neq \lambda_j)$$

then for $n \geq N$ and $k = k_*$

$$|(K_n K^*_{n+k+M})_{ij}| \leq C \exp(-n|\lambda_i - \lambda_j| + 2|r_i - r_j| n\varepsilon) \tag{10}$$

with $C = d \cdot \max(e^{-M\rho/2}, e^{-n\varepsilon})$. Thus if we choose $N \geq \log d/\varepsilon$ and $M \geq (2 \log d)/\rho$, then $C \leq 1$.

To establish (10), write

$$(K_n K^*_{n+k+M})_{ij} = \sum_l (K_n K^*_{n+M})_{il} (K_{n+M} K^*_{n+k+M})_{lj}.$$

We apply the estimate in Step III(a) to the term $(K_n K^*_{n+M})_{il}$ and the induction hypothesis to the term $(K_{n+M} K^*_{n+k+M})_{lj}$. By running through each of the four cases $\lambda_l = \lambda_i$, $\lambda_l = \lambda_j$, λ_l inside of (λ_i, λ_j) and λ_l outside of (λ_i, λ_j), we arrive at (10).

For example, if λ_l is inside of (λ_i, λ_j) we estimate

$$|(K_n K^*_{n+m})_{ij}| \leq e^{-n|\lambda_i - \lambda_j| + n\varepsilon}$$

$$\begin{aligned} |(K_{n+M} K^*_{n+k+M})_{lj}| &\leq C e^{-n|\lambda_l - \lambda_j| + 2|r_l - r_j| n\varepsilon} \\ &\leq C e^{-n\varepsilon} e^{-n|\lambda_l - \lambda_j| + 2|r_l - r_j| n\varepsilon}. \end{aligned}$$

Since $de^{-n\varepsilon} \leq 1$ and $|\lambda_i - \lambda_j| + |\lambda_l - \lambda_j| = |\lambda_i - \lambda_j|$, we arrive at (10). The other three cases are left to the reader. □

Step III(b) establishes Step III, since $|r_i - r_j|$ is always bounded by $d - 1$. In the proofs of Steps III(a) and III(b) we have considered only the case $\lambda_i, \lambda_j > -\infty$. The case where $\lambda_i = \lambda_j = -\infty$ is trivial, since $\lambda_i - \lambda_j$ is interpreted as 0. The remaining case, where one of λ_i, λ_j equals $-\infty$, is easy to arrive at by modifying the preceding proofs—and is also left to the reader.

Step IV

$$\lim_{n \to \infty} K^*_n D_n^{1/n} K_n = \Lambda$$

exists a.s.

PROOF. Let K, $\hat{K}$ be accumulation points of $\{K_n\}$. By Step I, we know that

$$D_n^{1/n} \to D = \operatorname{diag}(e^{\lambda_1}, \ldots, e^{\lambda_d}).$$

Thus it suffices to show that $K^*DK = \hat{K}^*D\hat{K}$, or equivalently $(\hat{K}K^*)D = D(\hat{K}K^*)$. But this follows from Step III, since $(\hat{K}K^*)_{ij} = 0$ whenever $\lambda_i \neq \lambda_j$. $\square$

Step V

Write

$$\lambda_1 = \cdots = \lambda_{i_1-1} < \lambda_{i_1} = \cdots = \lambda_{i_2-1} < \cdots < \lambda_{i_{r-1}} = \cdots = \lambda_d$$

and set $i_0 = 1$, $i_r = d + 1$. *Define*

$$V_k(n) = K_n^* \operatorname{span}\{\mathbf{e}_1, \ldots, \mathbf{e}_{i_k-1}\}, \qquad 1 \leq k \leq r$$

where $\mathbf{e}_i$ *are the standard unit vectors. Then*

$$V_k(n) \to V_k = K^* \operatorname{span}\{\mathbf{e}_1, \ldots, \mathbf{e}_{i_k-1}\}.$$

Proof. Again it suffices to show that

$$K^* \operatorname{span}\{\mathbf{e}_1, \ldots, \mathbf{e}_{i_k-1}\} = \hat{K}^* \operatorname{span}\{\mathbf{e}_1, \ldots, \mathbf{e}_{i_k-1}\}$$

and this amounts to showing that span $\{\mathbf{e}_1, \ldots, \mathbf{e}_{i_k-1}\}$ is invariant under $\hat{K}K^*$. Indeed if $\mathbf{u} \in \operatorname{span}\{\mathbf{e}_1, \ldots, \mathbf{e}_{i_k-1}\}$ then

$$\sum_{j=k}^{i_k-1} (\hat{K}K^*)_{ij} u_j = 0, \qquad \forall i \geq i_k$$

since $(\hat{K}K^*)_{ij} = 0$ whenever $\lambda_i \neq \lambda_j$. $\square$

Observe that $V_k(n)$ is the sum of the eigenspaces of $({}^nA^1)^*({}^nA^1)$ over eigenvalues $\leq \mu_{i_k-1}^2(n)$.

Step VI

If $\mathbf{v} \in V_k \setminus V_{k-1}$ *then*

$$\lim_{n\to\infty} \frac{1}{n} \log \|{}^nA^1\mathbf{v}\| = \lambda_k'.$$

Proof. Let $\mathbf{v} = K^*\mathbf{u}$ where $u_j = 0$ for $j \geq i_k$ and

$$\sum_{j=i_{k-1}}^{i_k-1} u_j^2 \neq 0.$$

Then

$$\|{}^nA^1\mathbf{v}\| = \|D_nK_nK^*\mathbf{u}\|.$$

By Step III

$$|\mu_i(n)(K_nK^*)_{ij}| \leq e^{n\lambda_j+n\varepsilon}$$

so that

$$\limsup_{n\to\infty} \frac{1}{n} \log \|{}^nA^1\mathbf{v}\| \le \lambda_{i_{k-1}} = \lambda'_k.$$

On the other hand, since $K_n K^*$ is orthogonal

$$\sum_{j=i_{k-1}}^{i_k-1} (K_n K^* u)_i^2 \to \sum_{j=i_{k-1}}^{i_k-1} u_j^2$$

so that

$$\liminf_{n\to\infty} \frac{1}{n} \log \|{}^nA^1\mathbf{v}\| \ge \lambda'_k. \qquad \square$$

Steps IV, V, and VI establish all the conclusions claimed in Oseledec's Theorem.

For an example of Oseledec's Theorem, consider the 2×2 triangular matrices

$$A_n = \begin{bmatrix} a_n & c_n \\ 0 & b_n \end{bmatrix}.$$

Let $a = \mathbb{E} \log |a_n|$, $b = \mathbb{E} \log |b_b|$. The A_ns have the invariant subspace

$$\Gamma = \operatorname{span}\left\{\begin{bmatrix} 1 \\ 0 \end{bmatrix}\right\}.$$

It can be checked that

$${}^nA^1 = \begin{bmatrix} a_n \dots a_1 & \sum_{k=1}^{n} a_n \dots a_{k+1} c_k b_{k-1} \dots b_1 \\ 0 & b_n \dots b_1 \end{bmatrix}.$$

If $a < b$ then all vectors of Γ grow like e^{na} (under repeated application of the As), but any vector $\mathbf{u} \notin \Gamma$ grows as e^{nb}. On the other hand, if $a > b$ then the growth rate e^{nb} is only attained for $\mathbf{u} = \begin{bmatrix} u_1 \\ u_2 \end{bmatrix}$ when

$$\frac{u_1}{u_2} = -\sum_{k=1}^{\infty} \frac{c_k}{a_k} \frac{b_{k-1}}{a_{k-1}} \cdots \frac{b_1}{a_1}, \tag{11}$$

and this defines a random subspace Γ'. Thus, in order to achieve the lower growth rate e^{nb}, the initial vector $\mathbf{u}$ must belong to Γ'. Thus either $\mathbf{u}$ has to be random itself, or else *typically* the growth rate e^{nb} will be achieved with probability zero. Moreover, note that in the former case the randomness is such that $\mathbf{u}$ would have to depend on the entire sequence (A_n), as seen from (11).

Oseledec's Theorem should properly be thought of as a result about *random dynamical systems*. It is concerned with the evolution of discrete-time processes

$$\mathbf{X}_n = A_n \mathbf{X}_{n-1}, \qquad n \geq 1. \tag{12}$$

It states that $\|\mathbf{X}_n\|$ grows like $e^{n\lambda_k}$ if $\mathbf{X}_0 = \mathbf{u} \in V_k - V_{k-1}$. The scalars λ_k and subspaces E_k constitute the "spectral decomposition" for the dynamical system (12). The constant λ from (6) is the same as λ_1, the largest "eigenvalue." Sometimes the λ_ks are called the *Lyapunov exponents* for the system (12), named after their counterparts in the non-random case.

If the A_ns are i.i.d., then (12) defines a Markov chain on $\mathbb{R}^d$. Its transition probabilities are

$$P(\mathbf{x}, B) = \mathbb{P}_{\mathbf{x}}(X_1 \in B) = \mathbb{P}(A\mathbf{x} \in B)$$

where B is a subset of $\mathbb{R}^d$, and A is a *random* $d \times d$ *matrix* with the (common) distribution of the A_ns. Oseledec's Theorem asserts that *(almost) all trajectories of this chain have an asymptotic geometric rate of growth.*

Fractals

A peculiar application of products of random matrices is to *fractal geometry*, in particular, to a *probabilistic algorithm for image generation.* The simplest form of the algorithm is illustrated in Figure 1. The leaf is generated as follows. Pick any point $\mathbf{X}_0 \in \mathbb{R}^2$. There are four *affine transformations* $T: \mathbf{x} \mapsto A\mathbf{x} + \mathbf{b}$ listed on top of this figure and four probabilities p_i underneath them. Choose one of these transformations at random, according to the probabilities p_i, say $T(k)$ is chosen, and apply it to $\mathbf{X}_0$, thereby obtaining $\mathbf{X}_1 = T(k)\mathbf{X}_0$. Then choose a transformation again at random, independent of the previous choice, and apply it to $\mathbf{X}_1$, thereby obtaining $\mathbf{X}_2$. Continue in this fashion, and plot the orbit $\{\mathbf{X}_n\}$. The result is the leaf shown. By tabulating the frequencies with which the points $\mathbf{X}_n$ fall into the various pixels of the graphics window, one can actually plot the empirical distribution $\frac{1}{n+1}\sum_{k=0}^{n} \delta_{\mathbf{X}_k}$ (where $\delta_{\mathbf{X}_k}$ is the point mass at $\mathbf{X}_k$), using a grey scale to convert statistical frequency to color. The darker portions of the leaf correspond to high probability density.

One framework for this process is in terms of image encoding and compression. The coefficients of the transformations $T(i)$ and the probabilities p_i represent an encoding of the leaf image, and the algorithm for generating the leaf is the decoding procedure. The most significant problem in this area is the inverse, or encoding problem, of finding the parameters to encode a given target image.

To develop the appropriate model for this algorithm, let $T: \mathbf{x} \mapsto A\mathbf{x} + \mathbf{b}$ be a *random affine transformation.* We denote by $A = A(T)$ the linear part of T, and by $\mathbf{b} = \mathbf{b}(T)$ the translational part of T. Let $\mathbf{X} \in \mathbb{R}^d$ be a

$$T(1)\colon \mathbf{x} \mapsto \begin{pmatrix} 0.8 & 0 \\ 0 & 0.8 \end{pmatrix}\mathbf{x} + \begin{pmatrix} 0.1 \\ 0.04 \end{pmatrix} \qquad T(2)\colon \mathbf{x} \mapsto \begin{pmatrix} 0.5 & 0 \\ 0 & 0.5 \end{pmatrix}\mathbf{x} + \begin{pmatrix} 0.25 \\ 0.4 \end{pmatrix}$$

$$T(3)\colon \mathbf{x} \mapsto \begin{pmatrix} 0.355 & -0.355 \\ 0.355 & 0.355 \end{pmatrix}\mathbf{x} + \begin{pmatrix} 0.266 \\ 0.078 \end{pmatrix}$$

$$T(4)\colon \mathbf{x} \mapsto \begin{pmatrix} 0.355 & 0.355 \\ -0.355 & 0.355 \end{pmatrix}\mathbf{x} + \begin{pmatrix} 0.378 \\ 0.434 \end{pmatrix}$$

$$p_1 = \frac{1}{2}, \qquad p_2 = p_3 = p_4 = \frac{1}{6}$$

Figure 1. Maple Leaf

This image corresponds to the stationary distribution of a 2-D Markov chain. (The window here is $0 \le x, y \le 1$.)

random variable independent of T. If $T\mathbf{X}$ has the same distribution as $\mathbf{X}$, then $\mathbf{X}$ is called *stationary* (with respect to T).

There is a *natural Markov chain* $\{\mathbf{X}_n\}_{n \ge 0}$ on $\mathbb{R}^d$ *associated with* T. Let $\{T_n\}_{n \ge 1}$ be an i.i.d. sequence of affine transformations $\mathbb{R}^d \to \mathbb{R}^d$, all distributed like T. Define

$$\mathbf{X}_n = T_n \mathbf{X}_{n-1}, \qquad n \ge 1. \tag{13}$$

Then, much like (12), $\{\mathbf{x}_n\}$ is a Markov chain with transition probabilities

$$\mathbb{P}_{\mathbf{x}}(X_1 \in B) = \mathbb{P}(T\mathbf{x} \in B)$$

In fact this is precisely the process $\{\mathbf{X}_n\}_{n \ge 0}$, described in the probabilistic algorithm, used to generate the leaf. There, T had a discrete distribution

$$\mathbb{P}(T = T(i)) = p_i, \qquad 1 \le i \le 4.$$

Although *algebraically* we are thinking of T as a (random) affine transformation, *probabilistically* it is nothing more than a multivariate random vector. Together the matrix and translational part of T have $d^2 + d$ entries, and so the distribution of T can be characterized by a $d^2 + d$-dimensional distribution function $F_T(t)$. Of course, we want to exploit the algebraic structure here as well as the probabilistic structure!

Suppose the random variable $\mathbf{X} \in \mathbb{R}^d$ is stationary (with respect to T). Let $C \subseteq \mathbb{R}^d$ be the support of $\mathbf{X}$,

$$\begin{aligned} C &= \bigcap \{K \subseteq \mathbb{R}^d \text{ closed: } \mathbb{P}(\mathbf{X} \in K) = 1\} \\ &= \{\mathbf{x} \in \mathbb{R}^d \colon \mathbb{P}(\mathbf{X} \in G) > 0 \text{ for any neighborhood } G \text{ of } \mathbf{x}\}. \end{aligned}$$

Observe that since $\mathbf{X}$ and T are independent,

$$\mathbb{P}(T\mathbf{X} \in B) = \int \mathbb{P}(\mathbf{X} \in t^{-1}B)\,dF_T(t). \tag{14}$$

If $T\mathbf{X}$ is to have the same distribution as $\mathbf{X}$, then it follows from (14) by setting $B = C$ that $\mathbb{P}(T^{-1}C \supseteq C) = 1$. In particular $\{t \in \mathbb{R}^{d^2+d}\colon t^{-1}C \supseteq C\} \supseteq H$ where $H = \operatorname{supp}(T) \subseteq \mathbb{R}^{d^2+d}$, since this set on the left is closed. Thus

$$C \supseteq \overline{\bigcup_{t \in H} tC}.$$

On the other hand, by setting $B = \overline{\bigcup_{t \in H} tC}$ in (14) we find that $\mathbb{P}(\mathbf{X} \in B) = 1$, so that $B \supseteq C$. Thus we conclude that C satisfies the *self-covering property*

$$C = \overline{\bigcup_{t \in H} tC}. \tag{15}$$

Observe in all of this that *we are implicitly identifying* $\mathbb{R}^{d^2+d}$ *with the space of affine transformations* $\mathbb{R}^d \to \mathbb{R}^d$. Otherwise, things like tC, $t^{-1}C$ would have no meaning. Here is where the algebra comes in with the probability—in considering events involving $T \in \mathbb{R}^{d^2+d}$ with the interpretation that T is really an affine map $T\colon \mathbb{R}^d \to \mathbb{R}^d$.

We need some definitions now, relating to the dynamics of T. Say that a *nonempty closed subset* $S \subseteq \mathbb{R}^d$ is (i) *invariant* (under H) if $\bigcup_{t \in H} tS \subseteq S$; (ii) *self-covering* (with respect to H) if $\overline{\bigcup_{t \in H} tS} = S$; and (iii) *minimal invariant* (with respect to H) if it is invariant, but no proper (closed) subset of it is invariant. Observe that minimal invariant sets are self-covering, but the converse does not necessarily hold.

One case where we can be sure of the existence and uniqueness of a minimal invariant set is when H contains a strictly contractive transformation t. In this case, every invariant set must contain the fixed point of

t, and so the intersection of *all* invariant sets, being nonempty, is the only minimal invariant set. Later we shall learn of a more general condition on H under which one can assert the uniqueness of minimal invariant sets.

If the transformations $t \in H$ are all uniformly strictly contractive and bounded, so that

$$\alpha = \sup_{t \in H} \|A(t)\| < 1, \qquad \beta = \sup_{t \in H} \|\mathbf{b}(t)\| < \infty, \tag{16}$$

then we can ensure the existence and uniqueness of a *compact* self-covering set. (N.B. The matrix norm is intended to be the operator norm, which corresponds to the vector norm.)

Lemma IV. *Under assumptions* (16) *there is exactly one nonempty compact subset* $\mathrm{C} \subseteq \mathbb{R}^{\mathrm{d}}$ *that is self-covering with respect to* H.

PROOF. To show the existence of C simply observe that the closed ball

$$\left\{\mathbf{x} \in \mathbb{R}^d \colon \|\mathbf{x}\| \leq \frac{\beta}{1-\alpha}\right\}$$

is invariant, and hence the intersection of all invariant sets must be bounded. To show the uniqueness of C we argue as follows. Suppose that the nonempty compact set $S \subseteq \mathbb{R}^d$ is also self-covering. Given $\varepsilon > 0$ for any $\mathbf{y} \in S$ there exist $t \in H$ and $\mathbf{x} \in S$ such that $\|\mathbf{y} - t\mathbf{x}\| \leq \varepsilon$. Thus

$$\begin{aligned} d(\mathbf{y}, C) &\leq d(\mathbf{y}, tC) \\ &\leq d(t\mathbf{x}, tC) + \varepsilon \leq \alpha d(\mathbf{x}, C) + \varepsilon \end{aligned}$$

where d is the distance function. Thus $d(S, C) \leq \alpha d(S, C)$ and since $d(S, C) < \infty$ and $\alpha < 1$ it follows that $d(S, C) = 0$, so that $S \subseteq C$. □

Sets in $\mathbb{R}^2$ can be thought of as black and white (i.e., binary) images. In this respect, self-covering sets in $\mathbb{R}^2$ are fractals, and the self-covering property (15) amounts to what Barnsley [1] calls the "collage," whereby C is covered by affine copies of itself. This is illustrated in Figures 2 and 3 where, in each case, H consists of the transformations listed. Observe in Figure 2 how C, the grey leaf, is covered by the four black leaves, each of which is an affine copy of C. This *collage property* is useful for encoding, since each of the black copies of C readily determines the corresponding affine transformation. It also shows that the support, or shape of the image only depends on the affine transformations in H, and not on the probabilities they get assigned. These probabilities only affect the coloration of the image. They can be used, say, to shift around the high- and low-density parts of the image.

Back to our original setting, suppose now that $T \in \mathbb{R}^{d^2+d}$ has support H. If $\mathbf{X} \in \mathbb{R}^d$ is stationary, then we see from (15) that C is self-covering

$$T(1)\colon \mathbf{x} \mapsto \begin{pmatrix} 0.8 & 0 \\ 0 & 0.8 \end{pmatrix} \mathbf{x} + \begin{pmatrix} 0.1 \\ 0.04 \end{pmatrix} \qquad T(2)\colon \mathbf{x} \mapsto \begin{pmatrix} 0.5 & 0 \\ 0 & 0.5 \end{pmatrix} \mathbf{x} + \begin{pmatrix} 0.25 \\ 0.4 \end{pmatrix}$$

$$T(3)\colon \mathbf{x} \mapsto \begin{pmatrix} 0.355 & -0.355 \\ 0.355 & 0.355 \end{pmatrix} \mathbf{x} + \begin{pmatrix} 0.266 \\ 0.078 \end{pmatrix}$$

$$T(4)\colon \mathbf{x} \mapsto \begin{pmatrix} 0.355 & 0.355 \\ -0.355 & 0.355 \end{pmatrix} \mathbf{x} + \begin{pmatrix} 0.378 \\ 0.434 \end{pmatrix}$$

$$p_1 = \frac{1}{2}, \qquad p_2 = p_3 = p_4 = \frac{1}{6}$$

Image under $T(1)$

Image under $T(2)$

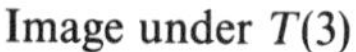

Image under $T(3)$

Image under $T(4)$

Figure 2. Self-Covering of the Maple Leaf
The black leaves form a collage of the grey background leaf.

(Text continues after color insert)

Random Dynamics for Generating Fractals

Listed in the tables which follow are two-dimensional affine maps $T(i)$ and corresponding probabilities p_i, for generating the fractal images in Figures 4–15. To generate these images from the data in the tables, carry out the following steps. Pick any point $\mathbf{X}_0 \in \mathbb{R}^2$ (say the origin). Choose one of the maps $T(i)$ at random, according to the probabilities p_i, say $T(k)$ is chosen, and apply it to $\mathbf{X}_0$, thereby obtaining $\mathbf{X}_1 = T(k)\mathbf{X}_0$. Then choose a transformation again at random, independent of the previous choice, and apply it to $\mathbf{X}_1$, thereby obtaining $\mathbf{X}_2$. Continue in this fashion, and plot the orbit $\{\mathbf{X}_n\}_{n \geq n_0}$, starting from a point $\mathbf{X}_{n_0}$ sufficiently along in the sequence. "Sufficiently along" means that $\mathbf{X}_{n_0}$ has already been drawn into the attractor. Typically $n_0 = 10$ suffices, but by choosing $\mathbf{X}_0$ to be the fixed point of one of the maps $T(i)$, one ensures that $\mathbf{X}_0$ itself is already in the attractor, in which case n_0 can be taken as zero—that is, plotting can begin at once with $\mathbf{X}_0$. By tabulating the frequencies with which the points $\mathbf{X}_n$ fall into the various pixels of the graphics window, one can use a color table to convert statistical frequency to color.

The images shown on the following pages were drawn using the GPLOT graphics package developed at the Pittsburgh Supercomputing Center. A pseudo-spectral red to blue 8-bit color table was used.

Figure 4

	a_{11}	a_{12}	a_{21}	a_{22}	b_1	b_2	p
Map # 1	0.833333	0.0	0.0	0.882353	0.083333403	0.1035696	0.655
Map # 2	−0.375	−0.4466915	0.074073769	−0.147059	0.7922795	0.2453809	0.08
Map # 3	−0.375	0.4466915	−0.074073769	−0.147059	0.5827206	0.3212355	0.08
Map # 4	−0.28764	−0.2797256	0.033957329	−0.086204	0.7002844	0.2775254	0.045
Map # 5	−0.115637	0.280683	−0.067051552	−0.093134999	0.5139341	0.329678	0.045
Map # 6	−0.139998	0.1583156	−0.04167822	−0.052908	0.5520328	0.3269508	0.02
Map # 7	−0.2	−0.079411507	0.1481484	−0.058823999	0.6122172	0.1297763	0.015
Map # 8	−0.2	0.079411507	0.1481484	0.058823999	0.5877829	0.1116766	0.015
Map # 9	−0.166667	−0.07762301	0.053332992	−0.074946001	0.5939498	0.2008248	0.015
Map #10	−0.193333	0.082976006	−0.123333	−0.049518	0.5813398	0.2648581	0.015
Map #11	0.213333	0.0	0.0	0.171306	0.3933331	0.1954045	0.015

For each of these figures the IFS maps are displayed as:

$$T: \begin{pmatrix} x \\ y \end{pmatrix} \rightarrow \begin{pmatrix} a_{11} & a_{12} \\ a_{21} & a_{22} \end{pmatrix} \begin{pmatrix} x \\ y \end{pmatrix} + \begin{pmatrix} b_1 \\ b_2 \end{pmatrix}$$

and their probabilities are denoted by p. The window is $0 \leq x,y \leq 1$.

Figure 5

	a_{11}	a_{12}	a_{21}	a_{22}	b_1	b_2	p
Map # 1	0.3823899	−0.0026093	0.0039139	0.1944659	0.4940943	0.3604451	0.06505
Map # 2	0.8512781	−0.4440292	0.4433769	0.8512783	0.24871243	−0.1157208	0.93495

Figure 6

	a_{11}	a_{12}	a_{21}	a_{22}	b_1	b_2	p
Map # 1	0.38	0.0	0.0	0.38	0.2630392	0.527633	0.12663
Map # 2	0.309017	−0.9510565	0.9510565	0.309017	0.7001897	−0.1105976	0.87337

Figure 7

	a_{11}	a_{12}	a_{21}	a_{22}	b_1	b_2	p
Map # 1	0.8062628	−0.1105484	0.0789497	0.7338552	0.1087847	−0.0277995	0.5562
Map # 2	−0.2974557	0.4421937	−0.3635353	−0.2524461	0.1751297	1.087646	0.21847
Map # 3	−0.0704501	−0.5882008	−0.4149449	−0.0117416	1.0436185	0.9501509	0.22533

Figure 8

	a_{11}	a_{12}	a_{21}	a_{22}	b_1	b_2	p
Map # 1	0.282	0.0	0.0	0.278	0.6003081	0.3473343	0.08969
Map # 2	0.7847452	−0.5701534	0.5701534	0.7847452	0.3304751	−0.1327902	0.91031

Figure 9

	a_{11}	a_{12}	a_{21}	a_{22}	b_1	b_2	p
Map # 1	0.4	0.0	0.0	0.4	0.2547549	0.5108053	0.13472
Map # 2	0.2575142	−0.9592138	0.9592138	0.2575142	0.7257136	−0.0921647	0.86528

Figure 10

	a_{11}	a_{12}	a_{21}	a_{22}	b_1	b_2	p
Map # 1	0.183953	0.0	0.0	0.1846053	0.7331354	0.4357292	0.06982
Map # 2	0.8728180	−0.4115444	0.4115444	0.8728180	0.2725137	−0.1435450	0.93018

Figure 11

	a_{11}	a_{12}	a_{21}	a_{22}	b_1	b_2	p
Map # 1	0.5	0.5	−0.5	0.5	0.125	0.625	0.5
Map # 2	0.5	0.5	−0.5	0.5	−0.125	0.375	0.5

Figure 12

	a_{11}	a_{12}	a_{21}	a_{22}	b_1	b_2	p
Map # 1	0.8	0.0	0.0	0.8	0.1	0.04	0.25
Map # 2	0.5	0.0	0.0	0.5	0.25	0.4	0.25
Map # 3	0.355	−0.355	0.355	0.355	0.266	0.078	0.25
Map # 4	0.355	0.355	−0.355	0.355	0.378	0.434	0.25

Figure 13

	a_{11}	a_{12}	a_{21}	a_{22}	b_1	b_2	p
Map # 1	0.7505016	−0.4076519	0.3990582	0.7458686	0.4007756	0.0257635	0.68136
Map # 2	−0.493151	−0.0000002	0.0000002	−0.516634	0.4842762	0.5452752	0.31864

Figure 14

	a_{11}	a_{12}	a_{21}	a_{22}	b_1	b_2	p
Map # 1	0.0	0.0	0.0	0.0	0.5	0.14	0.0025
Map # 2	0.0	0.0	0.0	0.0	0.716	0.284	0.0025
Map # 3	0.0	0.0	0.0	0.0	0.6272	0.3728	0.0025
Map # 4	0.0	0.0	0.0	0.0	0.5	0.32	0.0025
Map # 5	0.058196001	−0.991736	0.9919420	−0.01652899	0.9639354	0.014254302	0.495
Map # 6	0.058594	−0.988337	0.98543	−0.013659	0.9623597	0.018278122	0.495

Figure 15

	a_{11}	a_{12}	a_{21}	a_{22}	b_1	b_2	p
Map # 1	0.3358024	−0.2729940	0.1487280	0.2633952	0.39685055	0.5127947	0.05507
Map # 2	0.5080619	−0.8290816	0.8290816	0.5080619	0.5549108	−0.1635064	0.94493

4

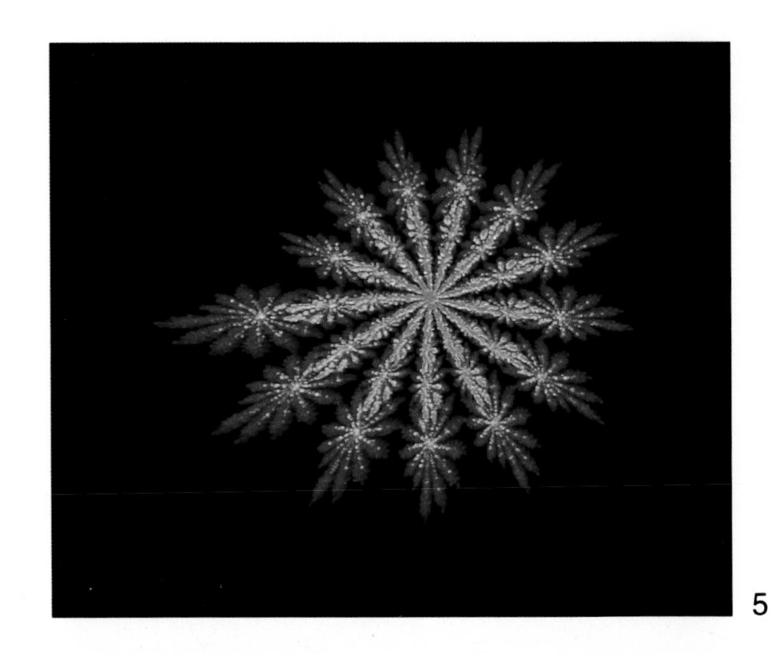

5

6

7

8

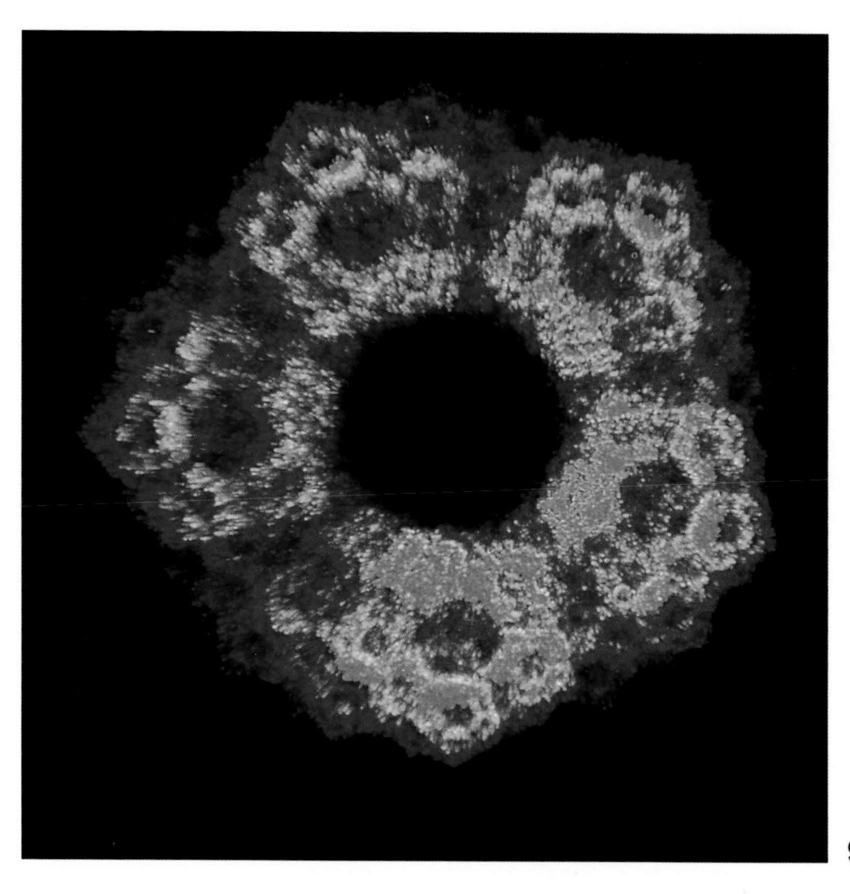

9

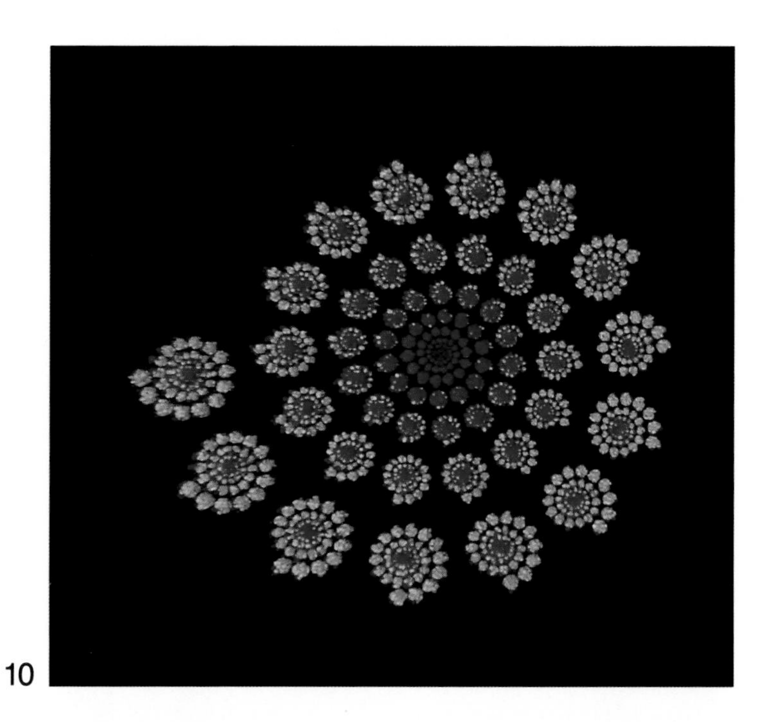

10

11

12

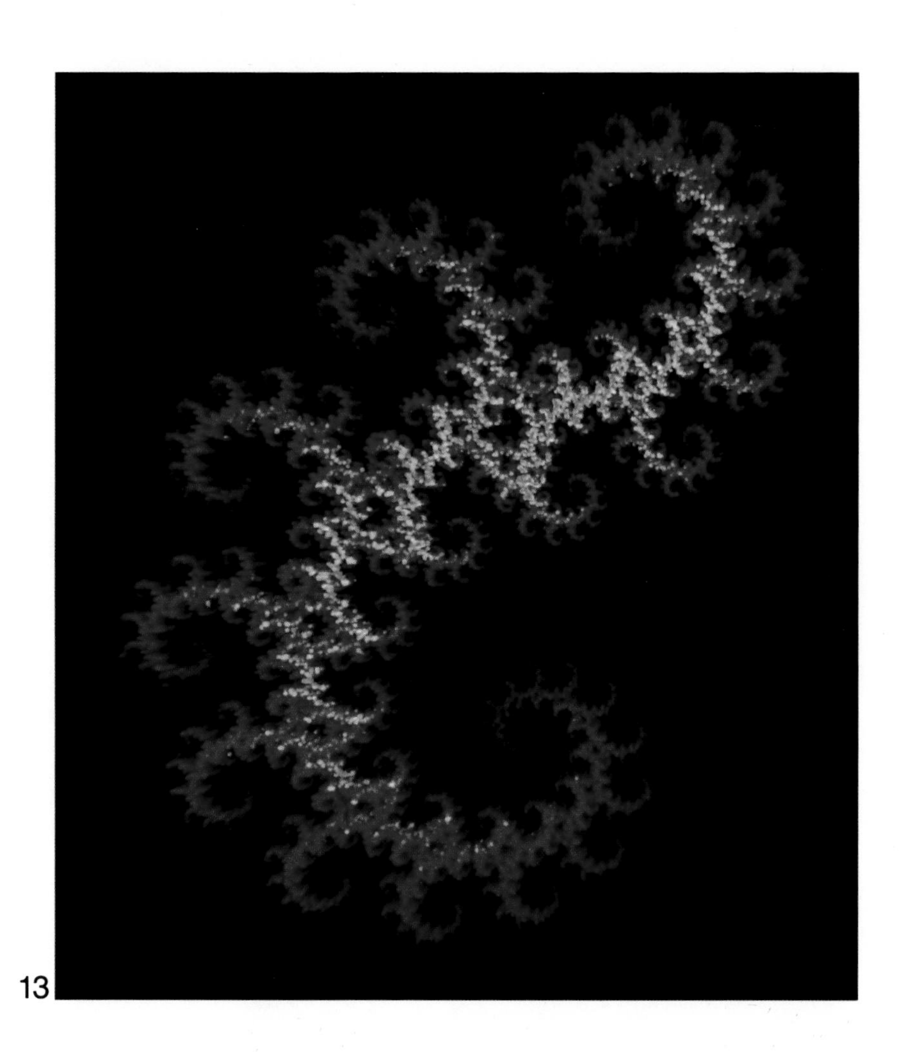

13

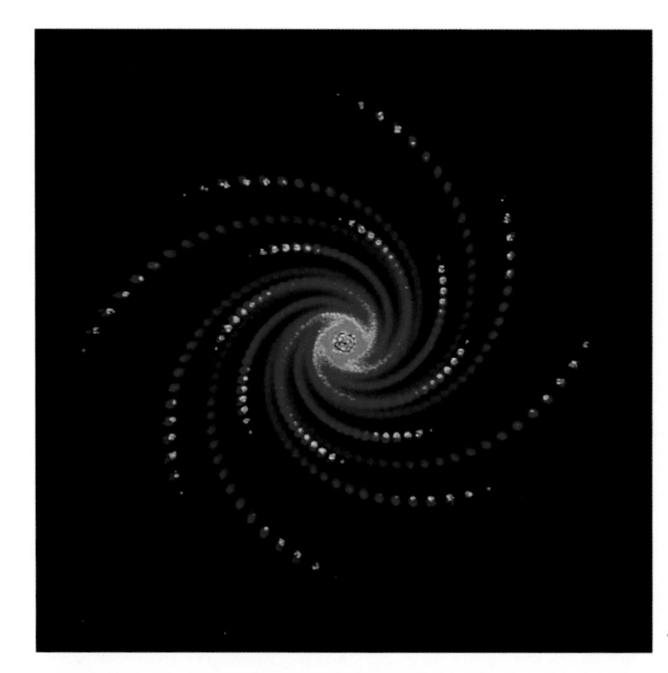

14

15

$$T(1)\colon \mathbf{x} \mapsto \begin{pmatrix} 0.5 & 0 \\ 0 & 0.5 \end{pmatrix} \mathbf{x} \qquad T(2)\colon \mathbf{x} \mapsto \begin{pmatrix} 0.5 & 0 \\ 0 & 0.5 \end{pmatrix} \mathbf{x} + \begin{pmatrix} 0.5 \\ 0 \end{pmatrix}$$

$$T(3)\colon \mathbf{x} \mapsto \begin{pmatrix} 0.5 & 0 \\ 0 & 0.5 \end{pmatrix} \mathbf{x} + \begin{pmatrix} 0.25 \\ 0.5 \end{pmatrix}$$

$$p_1 = p_2 = p_3 = \frac{1}{3}$$

A

F D

C E B

Figure 3. Self-Covering of the Sierpinski Triangle

Triangles ADF, BDE and CEF form a collage of triangle ABC. (The window here is $0 \le x, y \le 1$.)

(with respect to H). Our general interest lies in the following type of problem. Given an i.i.d. sequence $\{T_n\}_{n \ge 1}$ distributed like T, construct a (simple) stochastic process $\{\mathbf{X}_n\}_{n \ge 0}$ whose empirical distributions $\frac{1}{n+1} \sum_{k=0}^{n} \delta_{\mathbf{X}_k}$ converge weakly to a stationary random variable a.s. This is equivalent to requiring that with probability one the plot of the orbit of any *single* trajectory $\{\mathbf{X}_n\}_{n\geq 0}$ will produce the desired image. In this case there would be no need to ever run more than one trajectory. Of course we will be working in settings where there exist stationary random variables $\mathbf{X}$, and we will discuss this point first.

In order to invoke our results about products of random matrices, we make the following integrability assumption

$$\mathbb{E}[\log^+\|A\| + \log^+\|\mathbf{b}\|] < \infty \tag{17}$$

where $A = A(T), \mathbf{b} = \mathbf{b}(T)$. Let (A_n) be an i.i.d. sequence of random $d \times d$ matrices distributed like $A = A(T)$. On account of (17) the limit λ in (6) exists, $-\infty \leq \lambda < \infty$, and is constant a.s. Our next requirement is on λ:

$$\lambda < 0;$$

or equivalently

$$\mathbb{E}\log\|{}^nA^1\| < 0 \qquad \text{for some } n, \tag{18}$$

where ${}^nA^1 = A_n \dots A_1$. As mentioned earlier, this assumption guarantees that the iterates ${}^nA^1$ converge to zero exponentially fast as $n \to \infty$, a.s. We refer to this assumption as *average contractivity*. It certainly holds whenever the transformations in H are uniformly strictly contractive, but this is not a necessary condition.

Our next result shows that the Markov chain $\{\mathbf{X}_n\}_{n\geq 0}$ has the property we seek; namely, with probability one, the empirical distributions of any (single) trajectory converge weakly to a stationary random variable $\mathbf{X}$.

Theorem V. *Let* T: $\mathbb{R}^d \to \mathbb{R}^d$ *be a random affine transformation* $\mathbf{x} \mapsto \mathbf{A}\mathbf{x} + \mathbf{b}$, *and let* $\{\mathbf{X}_n\}$ *be the Markov chain in* $\mathbb{R}^d$ *defined by* (13). *Under assumptions* (17) and (18) *the following conclusions hold.*

(i) **Unique Stationary Distribution:** *There exists a unique random variable* $\mathbf{X} \in \mathbb{R}^d$, *up to distribution, that is stationary* (*with respect to* T).

(ii) **Asymptotic Stationarity:** *There holds* $\mathbf{X}_n \xrightarrow{\mathscr{D}} \mathbf{X}$.

(iii) **Law of Large Numbers:** *For any continuous function* f: $\mathbb{R}^d \to \mathbb{R}$ *with compact support*

$$\frac{1}{n+1}\sum_{k=0}^{n} f(\mathbf{X}_k) \to \mathbb{E}f(\mathbf{X}) \qquad \textit{a.s.}$$

The proof of this theorem relies on the following result.

Lemma VI. *Let* $\{\mathbf{X}_n\}_{n\geq 0}$ *be a Markov chain on* $\mathbb{R}^d$ *obeying the coupling condition*

$$\lim_{n\to\infty} (\mathbf{X}_n^{\mathbf{x}} - \mathbf{X}_n^{\mathbf{y}}) = 0 \qquad \text{a.s.} \tag{19}$$

for any $\mathbf{x}, \mathbf{y} \in \mathbb{R}^d$, *where* $\{\mathbf{X}_n^{\mathbf{x}}\}_{n\geq 0}$ *denotes the chain starting at* $\mathbf{x}$.

Suppose there is a d.f. F on $\mathbb{R}^d$ *such that the chain* $\{\mathbf{X}_n\}$ *with initial distribution* F *is stationary ergodic. Then this is the unique stationary distribution for the Markov chain, and for any continuous function* f: $\mathbb{R}^d \to \mathbb{R}$ *with compact support, and any* $\mathbf{x} \in \mathbb{R}^d$,

$$\frac{1}{n+1}\sum_{k=0}^{n} f(\mathbf{X}_k^{\mathbf{x}}) \to \int f(\mathbf{z})\, dF(\mathbf{z}) \qquad a.s. \tag{20}$$

Proof. Let $\{\mathbf{X}_n\}$ be the stationary ergodic chain in the statement of the lemma. It follows from the Pointwise Ergodic Theorem that

$$\lim_{n\to\infty}\frac{1}{n+1}\sum_{k=0}^{n} f(\mathbf{X}_k) = \int f(\mathbf{z})\, dF(\mathbf{z}) \qquad \text{a.s.}$$

It also follows from (19) that

$$\mathbb{P}\left(\lim_{n\to\infty}(\mathbf{X}_n^{\mathbf{x}} - \mathbf{X}_n) = 0\right)$$
$$= \int \mathbb{P}\left(\lim_{n\to\infty}(\mathbf{X}_n^{\mathbf{x}} - \mathbf{X}_n^{\mathbf{y}}) = 0\right) dF(\mathbf{y}) = 1.$$

Since f is *uniformly* continuous, we conclude that (20) holds. Indeed, given $\varepsilon > 0$, choose $\delta < 0$ so that

$$|f(\mathbf{y}) - f(\mathbf{z})| < \varepsilon/3 \qquad \text{whenever } \|\mathbf{y} - \mathbf{z}\| < \delta.$$

Then choose N_1 so that

$$\|\mathbf{X}_k^{\mathbf{x}} - \mathbf{X}_k\| < \delta \qquad \text{whenever } k > N_1;$$

and choose $N_2 \geq N_1$ so that

$$\frac{N_1 + 1}{N_2 + 1}\|f\| < \varepsilon/6$$

and

$$\left|\frac{1}{n+1}\sum_{k=0}^{n} f(\mathbf{X}_k) - \int f(\mathbf{z})\, dF(\mathbf{z})\right| < \varepsilon/3$$

whenever $n > N_2$. Whenever $n > N_2$, we can estimate

$$\left|\frac{1}{n+1}\sum_{k=0}^{n} f(\mathbf{X}_k^{\mathbf{x}}) - \int f(\mathbf{z})\, dF(\mathbf{z})\right|$$
$$\leq \frac{1}{n+1}\sum_{k=0}^{n} |f(\mathbf{X}_k^{\mathbf{x}}) - f(\mathbf{X}_k)| + \left|\frac{1}{n+1}\sum_{k=0}^{n} f(\mathbf{X}_k) - \int f(\mathbf{z})\, dF(\mathbf{z})\right|$$
$$< \frac{1}{N_2+1}\sum_{k=0}^{N_1} |f(\mathbf{X}_k^{\mathbf{x}}) - f(\mathbf{X}_k)| + \frac{1}{n+1}\sum_{k=N_1+1}^{n} |f(\mathbf{X}_k^{\mathbf{x}}) - f(\mathbf{X}_k)| + \varepsilon/3$$
$$< \varepsilon/3 + \varepsilon/3 + \varepsilon/3 = \varepsilon,$$

and we arrive at (20).

Suppose next that G is another stationary d.f. on $\mathbb{R}^d$ for $\{\mathbf{X}_n\}$. Then as earlier, it can be seen that

$$\lim_{n\to\infty} \frac{1}{n+1} \sum_{k=0}^{n} f(\mathbf{X}'_k) = \int f(\mathbf{z})\, dF(\mathbf{z}) \qquad \text{a.s.,} \tag{21}$$

where $\{\mathbf{X}'_n\}$ is the Markov chain with initial distribution G. Since $\mathbb{E}f(\mathbf{X}'_k) = \int f(\mathbf{z})\, dG(\mathbf{z})$ for each k, and since f is bounded, it follows using dominated convergence and taking expectations in (21) that

$$\int f(\mathbf{z})\, dG(\mathbf{z}) = \int f(\mathbf{z})\, dF(\mathbf{z}).$$

Since f is here an arbitrary continuous function with compact support, we get that $F = G$. $\square$

Proof of Theorem. Since

$$\mathbf{X}_n^{\mathbf{x}} - \mathbf{X}_n^{\mathbf{y}} = {}^nA^1(\mathbf{x} - \mathbf{y}),$$

assumption (18) guarantees that our chain $\{\mathbf{X}_n\}$ from (13) obeys the coupling condition (19). Thus, in order to prove the theorem we need to construct a d.f. that makes the chain stationary ergodic.

To this end, extend the *one-sided* i.i.d. sequence $\{T_n\colon n \geq 1\}$ to a *two-sided* i.i.d. sequence $\{T_n\colon n \in \mathbb{Z}\}$. This can be done by taking another one-sided i.i.d. sequence $\{T'_n\colon n \geq 1\}$, independent of our original sequence $\{T_n\colon n \geq 1\}$, and labeling its indices $0, -1, -2, \ldots$. Let $\mathbf{x} \in \mathbb{R}^d$ and consider a new process $\{\hat{\mathbf{X}}_n^{\mathbf{x}}\}_{n\geq 0}$ evolving as

$$\hat{\mathbf{X}}_n^{\mathbf{x}} = {}^0T^{-n+1}\mathbf{x}. \tag{22}$$

Recall that ${}^0T^{-n+1}$ denotes the product $T_0T_{-1}\cdots T_{-n+1}$. Observe that in distinction to (13), the successive transformations T_{-n} are applied from the *inside* here, rather than from the outside. The process $\{\hat{\mathbf{X}}_n^{\mathbf{x}}\}_{n\geq 0}$ is *no longer Markov*, in general, but it has two important properties.

(a) *For each fixed* n, $\mathbf{X}_n^{\mathbf{x}}$ *and* $\hat{\mathbf{X}}_n^{\mathbf{x}}$ *have the same distribution.*

(b) $\lim_{n\to\infty} \hat{\mathbf{X}}_n^{\mathbf{x}} = \mathbf{X}^{\mathbf{x}}$ *exists a.s.*

Property (a) is immediate, since both $\mathbf{X}_n^{\mathbf{x}}$ and $\hat{\mathbf{X}}_n^{\mathbf{x}}$ are given by a product of n i.i.d. copies of T acting on $\mathbf{x}$. To establish (b) one argues as follows. On account of the order in which the T_is are applied in (22),

$$\hat{\mathbf{X}}_{n+1}^{\mathbf{x}} - \hat{\mathbf{X}}_n^{\mathbf{x}} = {}^0A^{-n+1}(T_{-n}\mathbf{x} - \mathbf{x}).$$

Since $\mathbb{E}\log^+\|T_{-1}\mathbf{x} - \mathbf{x}\| < \infty$, by assumption (17), it follows from the Law of Large Numbers that

$$\lim_{n\to\infty} \frac{1}{n} \log^+\|T_{-n}\mathbf{x} - \mathbf{x}\| = 0 \qquad \text{a.s.} \tag{23}$$

Furthermore, by (18)

$$\lim_{n\to\infty} \frac{1}{n} \log\|{}^0A^{-n+1}\| < 0 \qquad \text{a.s.} \tag{24}$$

Together (23) and (24) imply that

$$\limsup_{n\to\infty} \frac{1}{n} \log\|\hat{\mathbf{X}}_{n+1}^{\mathbf{x}} - \hat{\mathbf{X}}_n^{\mathbf{x}}\| < \infty \qquad \text{a.s.}$$

so that $\sum_n \|\hat{\mathbf{X}}_{n+1}^{\mathbf{x}} - \hat{\mathbf{X}}_n^{\mathbf{x}}\| < \infty$ a.s. This establishes (b).

Observe further that for any $\mathbf{y} \in \mathbb{R}^d$

$$\hat{\mathbf{X}}_n^{\mathbf{x}} - \hat{\mathbf{X}}_n^{\mathbf{y}} = {}^0A^{-n+1}(\mathbf{x}-\mathbf{y}) \to 0 \quad \text{as} \quad n \to \infty \qquad \text{a.s.}$$

Thus for any $\mathbf{y} \in \mathbb{R}^d$, $\lim_{n\to\infty} \hat{\mathbf{X}}_n^{\mathbf{y}} = \mathbf{X}$ a.s., where $\mathbf{X} = \mathbf{X}^{\mathbf{x}}$. In other words, $\mathbf{X}^{\mathbf{x}}$ does not depend on $\mathbf{x}$ at all. On account of property (a), $\hat{\mathbf{X}}_n^{\mathbf{y}} \xrightarrow{\mathscr{D}} \mathbf{X}$ for any $\mathbf{y} \in \mathbb{R}^d$. This establishes the asymptotic stationarity in (ii).

The random variable $\mathbf{X}$ is invariant (with respect to T). Its d.f. $F_{\mathbf{X}}$ is therefore such that the chain $\{\mathbf{X}_n\}_{n\geq 0}$ in (13) with initial distribution $F_{\mathbf{X}}$ is stationary. Moreover if we take $\mathbf{X}_0 = \mathbf{X}$ then in fact $\{\mathbf{X}_n\}_{n\geq 0}$ will be ergodic as well. To see this, observe that since $\mathbf{X}_0 = \lim_{m\to\infty} T_0 T_{-1} \ldots T_{-m}\mathbf{x}$, it follows that each $\mathbf{X}_n$ is the *same* function

$$\mathbf{X}_n = u(T_n, T_{n-1}, \ldots);$$

namely, $\mathbf{X}_n = \lim_{m\to\infty} T_n T_{n-1} \ldots T_{n-m}\mathbf{x}$. So any event that is invariant for $\{\mathbf{X}_n\}_{n\geq 0}$ will be invariant for $\{T_n : n \in \mathbb{Z}\}$, and hence have probability zero or one. Thus the d.f. $F_{\mathbf{X}}$ is such that the chain $\{\mathbf{X}_n\}_{n\geq 0}$ is stationary ergodic, and the theorem follows from the lemma. □

Using the theory of convergence in distribution, as in Billingsley [3], conclusion (20) can be strengthened to all continuous *bounded* functions f. Indeed, since the space of continuous functions $f: \mathbb{R}^d \to \mathbb{R}$ with compact support is both *separable* and *convergence-determining*, it follows that

$$\frac{1}{n+1} \sum_{k=0}^{n} \delta_{\mathbf{X}_k} \xrightarrow{\mathscr{D}} F_{\mathbf{X}} \qquad \text{a.s.} \tag{25}$$

Then (20) follows for any continuous bounded function $f: \mathbb{R}^d \to \mathbb{R}$ by integrating both sides of (25) against $f(\mathbf{z})$. Thus we find that the chain $\{\mathbf{X}_n\}_{n\geq 0}$ satisfies the criterion we were after; namely, that its empirical distributions converge weakly to the stationary d.f., a.s.

On account of this convergence result (25), the closed set $A = \text{supp}(\mathbf{X}) \subseteq \mathbb{R}^d$ is called the *attractor* for the dynamical system (13). The term attractor is supposed to convey the dynamical sense in which the orbit $\{\mathbf{X}_n\}$ gets drawn toward the set A and then fills it in, no matter where it started.

This is in fact what one sees while watching the random algorithm for generating A. Moreover, if the orbit gets in A, then it stays in A from then on (since A is invariant under H). It follows from (25) that the orbit $\{\mathbf{X}_n\}$ is dense in A. The self-covering property (15) shows that the *shape* of A only depends on supp(T), i.e., on the transformations $T(i)$, $1 \leq i \leq N$, from the algorithm. The probabilities p_i affect the *coloration* of A, i.e., the statistics of the random variable $\mathbf{X}$ supported on A, but not the shape of A. So the p_is can be used to "shift around weight" on A. Then there are really two aspects to the random image algorithm. There is the *geometry*, which is depicted in the shape of A, and the *statistics* (or coloration), which is depicted in the invariant d.f. $F_{\mathbf{X}}$.

Corollary VII. *Under assumptions* (17) *and* (18) *there exists a unique minimal invariant set with respect to* H = supp(T), *namely*, C = supp($\mathbf{X}$).

PROOF. Simply observe that on account of the asymptotic stationarity in the theorem, every invariant set $S \subseteq \mathbb{R}^d$ with respect to H must contain C. For if we start the chain at $\mathbf{x} \in S$, then it remains in S forever. Thus $\mathbb{P}(\mathbf{X}_n \in S) = 1$, $\forall n$. Since $\mathbf{X}_n \xrightarrow{\mathscr{D}} \mathbf{X}$ and S is closed, we get that $\mathbb{P}(\mathbf{X} \in S) = 1$ also. □

Given a closed set H of affine transformations $t\colon \mathbb{R}^d \to \mathbb{R}^d$, we do not know of necessary and sufficient conditions for the existence of a random affine transformation T supported on H and satisfying (18), even in the case when H is a finite set. Clearly, though, a sufficient condition is that H contain a strictly contractive transformation t, since then we can arrange for (18) to be satisfied with $n = 1$ by choosing the distribution of T close to the atom δ_t.

Figures 4–15 show color images generated by the random algorithm described earlier, i.e., through products of i.i.d. affine transformations. In many cases there are only two affine transformations that the random transformation can take on (i.e., only two atoms).

Bibliographical Comments

The proof of the Pointwise Ergodic Theorem was adapted from Breiman [5], and the proof of the Mean Ergodic Theorem was adapted from Krengel [37]. Kingman's Theorem first appeared in [34] and was motivated by Hammersley and Welsh [24]. The survey article [35] discusses various applications.

One of the earliest works on products of random matrices, motivated by Bellman [2], is the article by Furstenberg and Kesten [21]. There it was shown that the leading characteristic exponent

$$\lambda = \lim_{n\to\infty} \frac{1}{n} \log\|{}^nA^1\|$$

exists a.s. In addition, under conditions of positivity the entry-wise limits

$$\lim_{n\to\infty} \frac{1}{n} \log({}^nA^1)_{ij}$$

were studied, and a CLT was proved for

$$\frac{\log({}^nA^1)_{ij} - n\lambda}{\sqrt{n}}.$$

Although Kingman's Subadditive Ergodic Theorem was not around yet at that time, the techniques in [21] resemble those used to prove Kingman's Theorem, especially the idea of examining

$$\lim_{N\to\infty} \frac{1}{N} \sum_{k=1}^{N} \mathbf{X}_{(k-1)r,\,kr}.$$

The works of Furstenberg [19] and [20] are the basic papers dealing with i.i.d. products in groups. Oseledec's work [43] developed the full spectrum, as presented here. The proof given here comes from Cohen et al. [9], which is in turn based on Raghunathan [47]. The part of the proof involving exterior products $\bigwedge^p A$ was taken from Krengel [37].

The example where A has normally distributed entries (an "isotropic" distribution), whereby AU is independent of U so that the LLN comes in directly, is due to Cohen and Newman [10]. See also the discussion in [42], for calculation of the full Oseledec spectrum.

The random algorithm for generating fractals is due to Barnsley [1]. He defines an *iterated function system with probabilities* to be a finite collection of (nonrandom) transformations $T(i)$, $1 \le i \le N$, with corresponding probabilities p_i. This amounts to a random transformation T with discrete distribution

$$\mathbb{P}(T = T(i)) = p_i, \qquad 1 \le i \le N.$$

Barnsley studies the Collage Theorem, the continuous dependence of the support image on the parameters of the $T(i)$s, the fractal dimension of the image, fractal interpolant curves, and many other fascinating topics. He has developed a compression board that uses the parameters of the $T(i)$s as a means of image compression, based on his algorithm for solving the inverse, or encoding problem.

Exercises

1. Consider the following algorithm

$x \leftarrow 0;$
for $n = 1{,}10000$
  plot x;
  choose $i \in \{0, 1\}$ randomly;
  $x \leftarrow \dfrac{x+i}{2};$
endfor.

(a) Consider the binary representation of a real number $x \in [0, 1]$

$$x = .b_1 b_2 \ldots \quad \Leftrightarrow \quad x = \frac{b_1}{2} + \frac{b_2}{4} + \frac{b_3}{8} + \cdots$$

(e.g., $\frac{1}{2} = .1000\ldots$, $\frac{7}{8} = .111000\ldots$, $\frac{9}{16} = .1001000\ldots$, etc.). Explain what the transformation $x \leftarrow \dfrac{x+i}{2}$ does, in terms of the bit sequence $.b_1 b_2 b_3 \ldots$, for $i = 0, 1$.

(b) Suppose the first 10 random choices of i are

$$1, 0, 0, 1, 0, 1, 1, 1, 0, 0.$$

Write down the bit sequence for the first 11 values of x.

(c) What is the output of this algorithm?

(d) What is the output of the following algorithm?

$x \leftarrow 0;$
for $n = 1{,}10000$
  plot x;
  choose $i \in \{0, 1, 2\}$ randomly;
  $x \leftarrow \dfrac{x+i}{3};$
endfor.

(e) What is the output of the following algorithm?

$x \leftarrow 0;$
for $n = 1{,}10000$
  plot x;
  choose $i \in \{0, 2\}$ randomly;
  $x \leftarrow \dfrac{x+i}{2};$
endfor.

(f) What is the output of the following algorithm?

$x \leftarrow 0;$
for $n = 1{,}10000$
  plot x;
  choose $i \in \{0, 1, 2, 3\}$ randomly;

$$x \leftarrow \frac{x+i}{3};$$

endfor.

2. For subsets $C \subseteq \mathbb{R}^d$ denote by conv C the convex hull of C, and by ext C the set of extreme points of conv C.

   (a) Prove that ext $C \subseteq \overline{C}$.
   (b) Prove that for subsets $C_1, \ldots, C_N \subseteq \mathbb{R}^d$

   $$\operatorname{conv} \bigcup_{i=1}^{N} C_i \supseteq \bigcup_{i=1}^{N} \operatorname{conv} C_i.$$

   (c) Let $T: \mathbb{R}^d \to \mathbb{R}^d$ be an affine transformation, and let $C \subseteq \mathbb{R}^d$. Prove that

   $$\operatorname{conv} TC = T(\operatorname{conv} C).$$

   (d) Let $T: \mathbb{R}^d \to \mathbb{R}^d$ be an affine transformation, and let $C \subseteq \mathbb{R}^d$ be compact. Prove that

   $$\operatorname{ext} TC \subseteq T(\operatorname{ext} C).$$

   (e) Suppose $C \subseteq \mathbb{R}^d$ satisfies the collage property

   $$C = \bigcup_{i=1}^{N} T_i C$$

   where $T_1, \ldots, T_N: \mathbb{R}^d \to \mathbb{R}^d$ are affine transformations. Prove that conv C is invariant with respect to $\{T_1, \ldots, T_N\}$.
   (f) Suppose C in part (e) is also compact. Prove that

   $$\operatorname{ext} C \subseteq \bigcup_{i=1}^{N} T_i (\operatorname{ext} C).$$

   (g) Explain the results in (e) and (f) geometrically, in terms of the attractor C.

3. Let $T: \mathbb{R}^d \to \mathbb{R}^d$ be a random affine transformation $\mathbf{x} \mapsto A\mathbf{x} + \mathbf{b}$, and let $\mathbf{X} \in \mathbb{R}^d$ be an invariant random variable with respect to T.

   (a) Find the condition satisfied by the characteristic function $\varphi_{\mathbf{X}}$, which corresponds to $\mathbf{X}$ being invariant.
   (b) Suppose A is nonrandom and strictly contractive. Find an expression for $\varphi_{\mathbf{X}}$.
   (c) Use the result from (b) to show that the invariant distribution is the d.f. of

   $$\sum_{n=0}^{\infty} A^n \mathbf{b}_{n+1},$$

   where $\{\mathbf{b}_n\}_{n \geq 1}$ is an i.i.d. sequence distributed like $\mathbf{b}$.
   (d) Generalize the result in (c) to the case where A is random.
   (e) Work out φ_X and then find the distribution of X, for the one-dimensional random affine transformation

   $$T: x \mapsto \frac{x+Y}{2},$$

   where Y is $\operatorname{bin}(n, \frac{1}{2})$.

(f) Is there a distribution for $T: \mathbb{R}^d \to \mathbb{R}^d$ satisfying (17) and (18) that will make $\mathbf{X} \in \mathbb{R}^d$ normally distributed?

4. Let $\{A_n\}_{n \geq 1}$ be an i.i.d. sequence of 2×2 matrices, and let $\{\mathbf{X}_n\}_{n \geq 0}$ be the 2-D Markov chain

$$\mathbf{X}_n = A_n \mathbf{X}_{n-1}, \qquad n \geq 1.$$

Let $M_n \in [-\infty, \infty]$ be the slope $M_n = X_{n,2}/X_{n,1}$.

(a) Show that $\{M_n\}_{n \geq 0}$ is also a Markov chain, and find its transition probabilities.
(b) Find a condition on the d.f. F_M, which makes it a stationary distribution for the slope process.
(c) Show that the Lyapunov exponent λ for $\{A_n\}_{n \geq 1}$ is given by

$$\lambda = \mathbb{E}\left[\log \left\| A \begin{pmatrix} 1 \\ M \end{pmatrix} \right\| - \log \left\| \begin{pmatrix} 1 \\ M \end{pmatrix} \right\| \right]$$

where A is distributed like the A_ns, and M is distributed like F_M, independent of A. Think about what happens if there is more than one stationary d.f. F_M.
(d) Show that the preceding expression for λ gives the same value for any operator norm $\|\cdot\|$.
(e) Let $T: \mathbb{R}^2 \to \mathbb{R}^2$ be a random affine transformation $\mathbf{x} \mapsto A\mathbf{x} + \mathbf{b}$. Affine transformations have the nice geometric property that they map parallel lines into parallel lines. Use this property to define a slope process $\{M_n\}_{n \geq 0}$ for an i.i.d. sequence of affine transformations $\{T_n\}$. Interpret this process by considering the dynamical evolution of lines under successive transformations T_n.
(f) Generalize the ideas in (a)–(e) to more than two dimensions.

5. Work out values for the stationary slope d.f. F_M (see Exercise 4), and find its support, in each of the following cases.

(a) $A = \begin{cases} \begin{pmatrix} 2 & 0 \\ 1 & 1 \end{pmatrix} & \text{w.p. } \frac{1}{2} \\ \begin{pmatrix} 3 & 1 \\ 1 & 3 \end{pmatrix} & \text{w.p. } \frac{1}{2} \end{cases}$

(b) $A = \begin{cases} \begin{pmatrix} 2 & 1 \\ 0 & 0 \end{pmatrix} & \text{w.p. } \frac{1}{2} \\ \begin{pmatrix} 1 & 1 \\ 1 & 1 \end{pmatrix} & \text{w.p. } \frac{1}{2} \end{cases}$

(c) $A = \begin{cases} \begin{pmatrix} 2 & 0 \\ 1 & 1 \end{pmatrix} & \text{w.p. } \frac{1}{2} \\ \begin{pmatrix} 2 & 1 \\ 0 & 1 \end{pmatrix} & \text{w.p. } \frac{1}{2} \end{cases}$

(d) $A = \begin{cases} \begin{pmatrix} 1 & -1 \\ 1 & 0 \end{pmatrix} & \text{w.p. } \frac{1}{2} \\ \begin{pmatrix} 0 & 1 \\ -1 & 1 \end{pmatrix} & \text{w.p. } \frac{1}{2} \end{cases}$

(e) $A = \begin{cases} \begin{pmatrix} 2 & 0 \\ 1 & 1 \end{pmatrix} & \text{w.p. } \frac{1}{2} \\ \begin{pmatrix} 2 & 0 \\ 0 & 1 \end{pmatrix} & \text{w.p. } \frac{1}{2} \end{cases}$

(f) $A = \begin{cases} \begin{pmatrix} 3 & 0 \\ 0 & 1 \end{pmatrix} & \text{w.p. } \frac{1}{2} \\ \begin{pmatrix} 3 & 0 \\ 2 & 1 \end{pmatrix} & \text{w.p. } \frac{1}{2} \end{cases}$

6. Let $\mathbf{Y} \in \mathbb{R}^d$ be a random variable, and let $T: \mathbb{R}^d \to \mathbb{R}^d$ be the transformation

$$T: \mathbf{x} \mapsto \frac{\mathbf{x} + \mathbf{Y}}{\alpha}$$

where $\alpha > 1$. Let $\mathbf{X} \in \mathbb{R}^d$ be an invariant random variable with respect to T. Prove that

$$\operatorname{supp}(\mathbf{X}) \subseteq \frac{1}{\alpha - 1} \operatorname{conv} \operatorname{supp}(\mathbf{Y}).$$

7. Find a random affine transformation $T: \mathbb{R}^2 \to \mathbb{R}^2$ taking on only two values $T(1)$ and $T(2)$ and satisfying (18), so that the attractor $A = \operatorname{supp}(\mathbf{X})$

   (a) consists of exactly N points (for any given $N \in \mathbb{N}$);
   (b) consists of the boundary of a circle;
   (c) consists of a solid square;
   (d) consists of a solid triangle;
   (e) consists of a spiral;
   (f) consists of the graph of the parabola $y = 4x(1 - x)$, $0 \le x \le 1$;
   (g) consists of the Koch snowflake curve shown here.

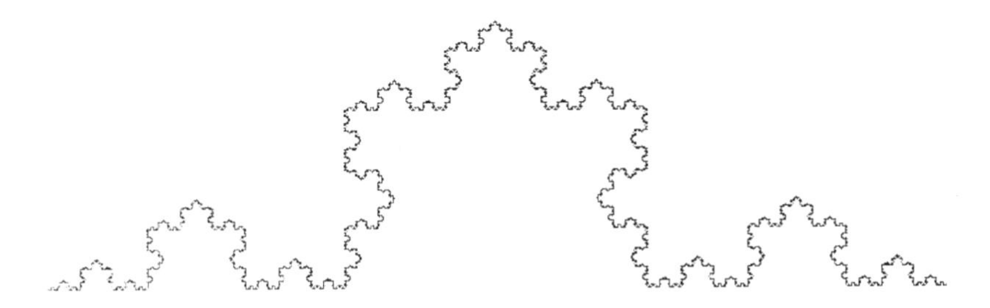

The Koch snowflake curve

8. Find a random affine transformation $T: \mathbb{R}^2 \to \mathbb{R}^2$ taking on only three values $T(1)$, $T(2)$, and $T(3)$ and satisfying (18), so that the attractor $A = \operatorname{supp}(\mathbf{X})$

   (a) consists of the boundary of a triangle;
   (b) consists of the boundary of a square;
   (c) consists of a solid circle;
   (d) consists of the boundary of a pentagon;
   (e) consists of a solid pentagon.

9. Let $T: \mathbb{R}^d \to \mathbb{R}^d$ be a random affine transformation $\mathbf{x} \mapsto A\mathbf{x} + \mathbf{b}$ satisfying (17) and (18); let $\mathbf{X} \in \mathbb{R}^d$ be an invariant random variable with respect to T.

   (a) Suppose we know all the moments of T up to order N; i.e., we know $\mathbb{E}p(a_{11}, a_{12}, \ldots, a_{dd}, b_1, \ldots, b_d)$ for any monomial p of degree $\le N$. Show how to obtain all the moments of $\mathbf{X}$ up to order N.
   (b) What do the formulas in (a) reduce to in the one-dimensional case $d = 1$?
   (c) Let $T: \mathbb{R} \to \mathbb{R}$ be the random affine transformation

$$x \mapsto \begin{cases} \frac{1}{3}x, & \text{w.p. } \frac{1}{2} \\ \frac{1}{3}x + \frac{2}{3}, & \text{w.p. } \frac{1}{2} \end{cases}$$

and let F_X be the invariant d.f. with respect to T. Show that F_X is the Cantor function constructed at the end of Section I, and compute all of its moments.

10. (M. Perrugia) Let $T: \mathbb{R} \to \mathbb{R}$ be the random affine map

$$x \mapsto \begin{cases} \frac{1}{2}x, & \text{w.p. } \frac{1}{2} \\ \frac{1}{2}x + \frac{1}{2}, & \text{w.p. } \frac{1}{2}. \end{cases}$$

(a) Let $\{X_n^1\}_{n \geq 0}$ and $\{X_n^2\}_{n \geq 0}$ be two independent Markov chains associated with T, evolving according to (13). Construct the new process $\{L_n = |X_n^1 - X_n^2|\}_{n \geq 0}$. Show that $\{L_n\}_{n \geq 0}$ is the Markov chain associated with a different random affine transformation T': $\mathbb{R} \to \mathbb{R}$, and find the distribution of T'.

(b) Use the technique of Exercise 9 to find the moments of the invariant random variable L (with respect to T'). Check your result by finding the distribution of L. Show why L corresponds to the distance between two independent uniformly distributed points in $[0, 1]$.

(c) Extend the results of (a) and (b) to the random affine transformation

$$x \mapsto \begin{cases} ax, & \text{w.p. } p \\ ax + (1 - a), & \text{w.p. } 1 - p \end{cases}$$

where $0 \leq a \leq \frac{1}{2}$, $0 < p < 1$. What is the expected distance between two independent points in $[0, 1]$, each distributed like the Cantor d.f. (in Exercise 9)?

(d) Let $\{X_n^1\}_{n \geq 0}, \ldots, \{X_n^d\}_{n \geq 0}$ be independent Markov chains associated with $T: \mathbb{R} \to \mathbb{R}$ given by

$$x \mapsto \begin{cases} \frac{1}{2}x, & \text{w.p. } \frac{1}{2} \\ \frac{1}{2}x + \frac{1}{2}, & \text{w.p. } \frac{1}{2}. \end{cases}$$

Let $\{\mathbf{X}_n = (X_n^{(1)}, \ldots, X_n^{(d)})\}_{n \geq 0}$ be the vector process obtained by rearranging $X_n^1, \ldots, X_n^d$ in increasing order (at each stage n). Show that $\{\mathbf{X}_n\}_{n \geq 0}$ is the Markov chain in $\mathbb{R}^d$ associated with a random affine transformation $\tilde{T}: \mathbb{R}^d \to \mathbb{R}^d$, and find the distribution of $\tilde{T}$. Let $\mathbf{X} \in \mathbb{R}^d$ be invariant with respect to $\tilde{T}$. What is the distribution of $\mathbf{X}$?

(e) Work out $\tilde{T}: \mathbb{R}^2 \to \mathbb{R}^2$ and the moments of $\mathbf{X}$ in (d), for the special case $d = 2$.

References

1. Barnsley, M.F., *Fractals Everywhere*, Academic Press, New York, 1988.
2. Bellman, R., Limit theorems for non-commutative operations I, *Duke Math. J.* 21 (1954), 491–500.
3. Billingsley, P., *Convergence of Probability Measures*, Wiley, New York, 1968.
4. Billingsley, P., *Probability and Measure*, Wiley, New York, 1979.
5. Breiman, L., *Probability*, Addison-Wesley, Reading, Mass., 1968.
6. Chung, K.L., *A Course in Probability Theory* (Second Ed.), Academic Press, New York, 1974.
7. Chung, K.L., *Markov Chains with Stationary Transition Probabilities* (Second Ed.), Springer-Verlag, New York, 1967.
8. Cinlar, E., *Introduction to Stochastic Processes*, Prentice-Hall, Englewood Cliffs, 1975.
9. Cohen, J.E., Kesten, H., and Newman, C.M., Oseledec's multiplicative ergodic theorem: A proof, in *Contemp. Math., Vol. 50: Products of Random Matrices*, J.E. Cohen et al., Eds., 1986, Amer. Math. Soc., Providence, R.I., 23–30.
10. Cohen, J.E. and Newman, C.M., The stability of large random matrices and their products, *Ann. Prob.* 12 (1984), 283–310.
11. Cox, D.R. and Miller, H.D., *The Theory of Stochastic Processes*, Wiley, New York, 1965.
12. DeGroot, M.H., *Probability and Statistics*, Addison-Wesley, Reading, 1975.
13. Doob, J.L., *Stochastic Processes*, Wiley, New York, 1953.
14. Dynkin, E.B., *Markov Processes–I & II*, Academic Press, New York, 1965 (translated from the Russian ed. of Springer-Verlag).
15. Ellis, R.S., *Entropy, Large Deviations and Statistical Mechanics*, Springer-Verlag, New York, 1985.
16. Ethier and Kurtz, T., *Markov Processes: Characterization and Convergence*, Wiley, New York, 1986.
17. Feller, W., *An Introduction to Probability Theory and its Applications*, Vols. I–II, Wiley, New York, 1968 (Vol. I, Third Ed., Revised Printing), 1971 (Vol. II, Second Ed.).
18. Freedman, D., *Markov Chains*, Holden-Day, San Francisco, 1971.

19. Furstenberg, H., Noncommuting random products, *Trans. Amer. Math. Soc.* 108 (1963), 377–428.
20. Furstenberg, H., A Poisson formula for semi-simple Lie groups, *Ann. Math.* 77 (1963), 335–386.
21. Furstenberg, H., and Kesten, H., Products of random matrices, *Ann. Math. Stat.* 31 (1960), 457–469.
22. Gihman and Skorohod, A.V., *The Theory of Stochastic Processes*, Vols. I–II, Springer-Verlag, New York, 1974 (Vol. I), 1975 (Vol. II), 1979 (Vol. III) (translated from the Russian eds.).
23. Gnedenko, B.V., *The Theory of Probability* (Fourth Ed.), Chelsea, New York, 1968 (translated from the Russian ed.).
24. Hammersley, J.M., and Welsh, J.A.D., First-passage percolation, subadditive processes, stochastic networks and generalized renewal theory, *Bernoulli-Bayes-Laplace Anniversary Volume*, J. Neyman and L.M. LeCam, Eds., Springer-Verlag, Berlin, 1965.
25. Harris, T.E., *The Theory of Branching Processes*, Springer-Verlag, New York, 1963.
26. Hille, E. and Phillips, R.S., *Functional Analysis and Semi-Groups* (Revised Ed.), Amer. Math. Soc. Colloq. Publ. Vol. XXXI, Providence, 1957.
27. Hoel, P.G., Port, S.C. and Stone, C.J., *Introduction to Probability Theory*, Houghton Mifflin, Boston, 1971.
28. Hoel, P.G., Port, S.C. and Stone, C.J., *Introduction to Stochastic Processes*, Houghton Mifflin, Boston, 1972.
29. Hogg, R.V. and Craig, A.T., *Introduction to Mathematical Statistics* (Fourth Ed.), Macmillan, New York, 1978.
30. Hogg, R.V. and Tanis, E.A., *Probability and Statistical Inference* (Second Ed.), Macmillan, New York, 1983.
31. Karlin, S. and Taylor, H.M., *A First Course in Stochastic Processes* (Second Ed.), Academic Press, New York, 1975.
32. Kemeny, J.G., Snell, J.L. and Knapp, A.W., *Denumerable Markov Chains*, Van Nostrand, Princeton, 1966.
33. Kendall, M. and Stuart, A., *The Advanced Theory of Statistics, Vol. I: Distribution Theory* (Fifth Ed.), Charles Griffin & Company, London, 1987.
34. Kingman, J.F.C., The ergodic theory of subadditive stochastic processes, *J. Roy. Stat. Soc. Ser. B* 30 (1968), 499–510.
35. Kingman, J.F.C., Subadditive ergodic theory, *Ann. Prob.* 1 (1973), 883–909.
36. Kolmogorov, A.N., *Foundations of the Theory of Probability*, Chelsea, New York, 1956 (translated from the Russian ed.).
37. Krengel, U., *Ergodic Theorems*, de Gruyter, New York, 1985.
38. Lamperti, J., *Probability*, W.A. Benjamin, Inc., New York, 1966.
39. Leadbetter, M.R., Lindgren, G. and Rootzén, H., *Extremes and Related Properties of Random Sequences and Processes*, Springer-Verlag, New York, 1983.
40. Loève, M., *Probability Theory Vols. I–II* (Fourth Eds.), Springer-Verlag, New York, 1977 (Vol. I), 1978 (Vol. II).
41. Mandl, P., *Analytical Treatment of One-Dimensional Markov Processes*, Springer-Verlag, New York, 1968.
42. Newman, C.M., The distribution of Lyapunov exponents: exact results for random matrices, *Comm. Math. Phys.* 103 (1986), 121–126.
43. Oseledec, V.I., A multiplicative ergodic theorem, Lyapunov characteristic numbers for dynamical systems, *Trans. Moscow Math. Soc.* 19 (1968), 197–221.
44. Parzen, E., *Modern Probability Theory and its Applications*, Wiley, New York, 1967.
45. Parzen, E., *Stochastic Processes*, Holden-Day, San Francisco, 1967.

46. Pazy, A., *Semigroups of Linear Operators and Applications to Partial Differential Equations*, Springer-Verlag, New York, 1983.
47. Raghunathan, M.S., A proof of Oseledec's multiplicative ergodic theorem, *Israel J. Math.* 32 (1979), 356–362.
48. Rényi, A., *Foundations of Probability*, Holden-Day, San Francisco, 1970.
49. Revuz, D., *Markov Chains*, North-Holland, New York, 1975.
50. Rockafellar, *Convex Analysis*, Princeton Univ. Press, Princeton, 1970.
51. Rogers, L.C.G. and Williams, D., *Diffusions, Markov Processes, and Martingales, Vol. 2: Itô Calculus*, Wiley, Chichester, 1987.
52. Romanovsky, V.I., *Discrete Markov Chains*, Wolters-Noordhoff, Groninger, 1970 (translated from the Russian ed.).
53. Rosenblatt, M., *Markov Processes: Structure and Asymptotic Behavior*, Springer-Verlag, New York, 1971.
54. Ross, S.M., *Introduction to Probability Models* (Third Ed.), Academic Press, New York, 1985.
55. Stroock, D.W., *An Introduction to the Theory of Large Deviations*, Springer-Verlag, New York, 1984.
56. Takacs, L. *Introduction to the Theory of Queues*, Oxford Univ. Press, New York, 1962.
57. Varadhan, S.R.S., *Large Deviations and Applications*, SIAM CBMS 46, Philadelphia, 1984.

Solutions (Sections I–V)

Section I

1. X is *binomial* with parameters $n = 2000$, $p = 1 - (1 - 0.6)^2 = 0.84$. Thus $\mu = np = 1680$ and $\sigma^2 = npq = 268.8$.

2. (i) N_n is *geometric* with parameter $p = \dfrac{1}{n}$. Thus $\mu = 1/p = n$ and $\sigma^2 = q/p^2 = n(n-1)$.
   (ii) Since the right key is equally likely to turn up on any of trials $1, \ldots, n$ it follows that N_n is *uniformly* distributed on $\{1, \ldots, n\}$. Thus $\mu = \dfrac{n+1}{2}$ and $\sigma^2 = \dfrac{n^2 - 1}{12}$.

3. In each case g is the inverse c.d.f. of the required distribution.
   (a) $g(u) = \tan \pi(u - \frac{1}{2})$.
   (b) $g(u) = -\dfrac{1}{\mu} \log(1 - u)$.
   (c) $g(u) = m + \sigma\Phi^{-1}(u)$, where Φ is the error function.

4. (a) Use the identities $\mathbb{E} \cos \pi X = \operatorname{Re} \varphi_X(\pi)$ and $\mathbb{E} \cos^2 \pi X = \frac{1}{2} + \frac{1}{2} \operatorname{Re} \varphi_X(2\pi)$.

$$\mathbb{E} \cos \pi X = \cos \pi m \, e^{-(1/2)\sigma^2\pi^2}$$

$$\operatorname{Var}(\cos \pi X) = \tfrac{1}{2}(1 - e^{-\sigma^2\pi^2})(1 - \cos 2\pi m \, e^{-\sigma^2\pi^2}).$$

   (b) Here, since X is integer-valued, $\cos \pi X = (-1)^X$ and $\cos^2 \pi X \equiv 1$. Thus

$$\mathbb{E} \cos \pi X = e^{-\lambda} \sum_{k=0}^{\infty} \frac{(-\lambda)^k}{k!} = e^{-2\lambda},$$

$$\operatorname{Var}(\cos \pi X) = 1 - e^{-4\lambda}.$$

(c) $$\mathbb{E}\cos \pi X = \frac{1}{2}\int_{-1}^{1} \cos \pi x \, dx = 0$$

$$\operatorname{Var}(\cos \pi X) = \frac{1}{2}\int_{-1}^{1} \cos^2 \pi x \, dx = \frac{1}{2}.$$

5. (a) This is *negative binomial* with parameters $r = 1$, $p = \dfrac{1}{1+a}$.
Thus $\mu = q/p = a$ and $\sigma^2 = q/p^2 = a(1+a)$.

(b) This is *negative binomial* with parameters $r = 1/\beta$, $p = \dfrac{1}{1+\alpha\beta}$.
Thus $\mu = r(q/p) = \alpha$ and $\sigma^2 = r(q/p^2) = \alpha(1+\alpha\beta)$.

(c) $Y = X - a$ has the density $f_Y(y) = \dfrac{1}{2\alpha} e^{-|y|/\alpha}$, and

$$\varphi_Y(u) = \frac{1}{1+\alpha^2 u^2} = 1 - \alpha^2 u^2 + \cdots.$$

Thus $\mu_X = a$ and $\sigma_X^2 = \sigma_Y^2 = -\varphi_Y''(0) = 2\alpha^2$.

(d) $X = e^Z$ where Z is $N(\alpha, \beta)$. Thus $\mu = e^{\alpha+(1/2)\beta^2}$, $\sigma^2 = e^{2\alpha+\beta^2}(e^{\beta^2} - 1)$.

6. (a) $Y = 2\left(\dfrac{X}{\alpha}\right)^2$ is $\chi^2(3)$. Thus $\mathbb{E}X^2 = \frac{3}{2}\alpha^2$ and $\mathbb{E}X^4 = \frac{15}{4}\alpha^4$. Since

$$\int_0^\infty x^3 e^{-x^2}\, dx = \frac{1}{2}\int_0^\infty x^2 d(-e^{-x^2}) = \int_0^\infty x e^{-x^2} dx = \frac{1}{2},$$

we also have $\mathbb{E}X = \dfrac{2\alpha}{\sqrt{\pi}}$. Thus

$$\operatorname{Var} X = \left(\frac{3}{2} - \frac{4}{\pi}\right)\alpha^2,$$

$$\mathbb{E}(\tfrac{1}{2}mX^2) = \tfrac{3}{4}m\alpha^2, \qquad \operatorname{Var}(\tfrac{1}{2}mX^2) = \tfrac{3}{8}m^2\alpha^4.$$

(b) $$\mathbb{E}|X - x_0| = 2\sqrt{\frac{Dt}{\pi}}(1 + e^{-x_0^2/Dt} - e^{-x_0^2/4Dt}) + 2x_0\left[2\Phi\left(\frac{2x_0}{\sqrt{2Dt}}\right) - \Phi\left(\frac{x_0}{\sqrt{2Dt}}\right) - 1\right]$$

$$\operatorname{Var}(|X - x_0|) = 2Dt - 4\sqrt{\frac{Dt}{\pi}}x_0 e^{-x_0^2/4Dt} + 4x_0^2\left[1 - \Phi\left(\frac{x_0}{\sqrt{2Dt}}\right)\right] - (\mathbb{E}|X - x_0|)^2.$$

7. (a) $\mathbb{E}|X - m| = \sigma\sqrt{\dfrac{2}{\pi}}\displaystyle\int_0^\infty x e^{-(1/2)x^2}\, dx = \sigma\sqrt{\frac{2}{\pi}}.$

(b) Use the centered moments from the notes:

$$\mathbb{E}X^3 = n^3p^3 + 3n^2p^2q + npq(q - p)$$

$$\mathbb{E}X^4 = n^4p^4 + 6n^3p^3q + n^2p^2q(7q - 4p) + npq(1 - 6pq)$$

To compute $\mathbb{E}|X - \mu|$ show by induction on m that

$$\sum_{k=0}^{m-1} (np - k)\binom{n}{k} p^k q^{n-k} = qm\binom{n}{m} p^m q^{n-m}, \qquad 1 \le m \le n.$$

Thus $\mathbb{E}|X - \mu| = 2qm\binom{n}{m} p^m q^{n-m}$ where $m = [np] + 1$.

8. (a) Use the inversion formula for φ_1, φ_3 to compute φ_2, φ_4,

$$\varphi_1(u) = \frac{a^2}{a^2 + u^2}; \qquad \varphi_2(u) = e^{-a|u|}; \qquad \varphi_3(u) = 2\frac{1 - \cos au}{\alpha^2 u^2};$$

$$\varphi_4(u) = \begin{cases} \dfrac{a - |u|}{a}; & |u| \le a, \\ 0, & |u| > a. \end{cases}$$

(b) Read off p_1, p_2 from the Fourier series:

$$\cos u = \tfrac{1}{2}e^{iu} + \tfrac{1}{2}e^{-iu}; \qquad p_1(1) = p_1(-1) = \tfrac{1}{2};$$

$$\cos^2 u = \tfrac{1}{2} + \tfrac{1}{4}e^{-2iu} + \tfrac{1}{4}e^{2iu};$$

$$p_2(2) = p_2(-2) = \tfrac{1}{4}, \qquad p_2(0) = \tfrac{1}{2}.$$

For φ_3, observe that $-X$ is *exponential* with parameter a; thus

$$f_3(x) = ae^{ax}, \qquad x \le 0.$$

φ_4 corresponds to the *uniform* distribution, since $\sin au = \tfrac{1}{2}(e^{iau} - e^{-iau})$:

$$f_4(x) = \frac{1}{2a}, \qquad |x| \le a.$$

Section II

1. Conditioned on X, Y is normal with

$$\mathbb{E}(Y|X) = \mu_Y + \rho\frac{\sigma_Y}{\sigma_X}(X - \mu_X), \qquad \operatorname{Var}(Y|X) = \sigma_Y^2(1 - \rho^2)$$

$$\mathbb{E}(e^{iuY}|X) = \exp\left\{i\left[\mu_Y + \rho\frac{\sigma_Y}{\sigma_X}(X - \mu_X)\right]u - \tfrac{1}{2}\sigma_Y^2(1 - \rho^2)u^2\right\}.$$

2. For each x, $\mathbb{E}(\cdot|X = x)$ is an expectation operator, so it follows from Jensen's inequality that

$$\varphi(\mathbb{E}(Y|X = x)) \le \mathbb{E}(\varphi(Y|X = x))$$

3. $\mathbb{P}(A|X_1) = \mathbb{P}(A|X_2) = \frac{1}{2}I_A + \frac{1}{2}I_{-A}$

$\mathbb{E}(Y|X_1) = \mathbb{E}(Y|X_2) = \frac{1}{2}Y(x) + \frac{1}{2}Y(-x).$

4. (a) $\mathbb{E}(Y|X) = \frac{1}{2}\frac{5 - 4X}{4 - 3X}.$

(b) $\mathbb{E}(Y|X) = \dfrac{2X - 6 + (X^2 + 4X + 6)e^{-X}}{X - 2 + (X + 2)e^{-X}}.$

(c) $\mathbb{E}(Y|X) = |X| + 2 + \dfrac{1}{|X| + 1}.$

(d) $\mathbb{E}(Y|X) = -\frac{1}{2}X.$

5. $f_X(x) = \dfrac{4\alpha(\alpha + 1)}{\beta^2} x \left(\dfrac{\beta}{\beta + 2x}\right)^{\alpha+2}.$

6. binomial $\left(X + Y, \dfrac{\lambda_X}{\lambda_X + \lambda_Y}\right).$

7. (a) Define states

$$s(A) = \begin{cases} 1, & a \text{ just won} \\ 0, & a \text{ just lost} \\ -1, & a \text{ just sat out} \end{cases}$$

$$\left.\begin{aligned} \mathbb{P}(A|s(A) = 1) &= \tfrac{1}{2} + \tfrac{1}{2}\mathbb{P}(A|s(A) = 0) \\ \mathbb{P}(A|s(A) = 0) &= \tfrac{1}{2}\mathbb{P}(A|s(A) = -1) \\ \mathbb{P}(A|s(A) = -1) &= \tfrac{1}{2}\mathbb{P}(A|s(A) = 1) \end{aligned}\right\} \Rightarrow \begin{aligned} &= \tfrac{4}{7} \\ &= \tfrac{1}{7} \\ &= \tfrac{2}{7}. \end{aligned}$$

Thus $\mathbb{P}(A) = \mathbb{P}(B) = \frac{5}{14}$, $\mathbb{P}(C) = \frac{4}{14}$.

(b) $$\left.\begin{aligned} \mathbb{E}(T|s(A) = 1) &= 1 + \tfrac{1}{2}\mathbb{E}(T|s(A) = 0) \\ \mathbb{E}(T|s(A) = 0) &= 1 + \tfrac{1}{2}\mathbb{E}(T|s(A) = -1) \\ \mathbb{E}(T|s(A) = -1) &= 1 + \tfrac{1}{2}\mathbb{E}(T|s(A) = 1) \end{aligned}\right\} \Rightarrow \begin{aligned} &= 2 \\ &= 2 \\ &= 2. \end{aligned}$$

Thus $\mathbb{E}T = 3$.

8. $\mathbb{E}T = 4$.

9. $$\mathbb{E}(T|T \wedge t) = \begin{cases} T, & T \wedge t < t \\ t + 1, & T \wedge t = t \end{cases}$$

$$\mathbb{E}(T|T \vee t) = \begin{cases} T, & T \vee t > t \\ 1 - \dfrac{t}{e^t - 1}, & T \vee t = t. \end{cases}$$

10. The joint *pdf* $f_{XY}(x, y)$ is independent of y, $1 \le y \le x$. Set $Z = Y - X$. Then $f_{XY}(x, y) = f_Y(y)f_Z(x - y)$. Thus

$$(*) \qquad f_Y(y)f_Z(x - y) = f_Z(0)f_Y(x) = f_Y(0)f_Z(x), \qquad 1 \le y \le x.$$

Integrate the first equality in (∗) from $x = y$ to $x = \infty$. Then

$$f_Y(y) = f_Z(0) \int_y^\infty f_Y(x)\, dx,$$

from which follows that $f_Y(y) = \beta e^{-\beta y}$ where $\beta = f_Z(0)$. Now use the second equality in (∗) to conclude that $f_Z = f_Y$.

11. Write $X = Z_1 + \cdots + Z_m$ where the Z_is are *binomial* with parameters n and $p = 1/2$. Thus

$$\mathbb{E}X = m\mathbb{E}Z_1 = \frac{mn}{2}.$$

The Z_is are not independent, so we expand

$$\begin{aligned}\operatorname{Var} X &= \sum_{i=1}^m \operatorname{Var}(Z_i) + \sum\sum_{i \neq j} \operatorname{Cov}(Z_i, Z_j)\\ &= m \operatorname{Var}(Z_1) + m(m-1) \operatorname{Cov}(Z_1, Z_2).\end{aligned}$$

The expression for Var X is thus quadratic in m with no zero-order term. Since Var $X = 0$ when $m = 2^n$ (for then X is nonrandom), we must have Var $X = Cm(2^n - m)$ for same constant C. Finally, since Var X = Var $Z_1 = \frac{1}{4}n$ when $m = 1$, this constant is given by $C = \dfrac{n}{4(2^n - 1)}$; and thus

$$\operatorname{Var} X = \frac{mn}{4} \frac{2^n - m}{2^n - 1}.$$

12. X is *Poisson*(λY), where Y itself is *Poisson*(μ).

$$\mathbb{P}(X = 0) = e^{-\mu(1-e^{-\lambda})}$$

$$\mathbb{E}X = \mu\lambda, \qquad \operatorname{Var} X = \mu\lambda(\lambda + 1)$$

13. $\mathbb{P}(|X - Y| \le 10) = \frac{11}{36}$.

14. $\dfrac{X^2 + Y^2}{9}$ is $\chi^2(2)$, $\mathbb{P}(X^2 + Y^2 \le 9) = 1 - e^{-1/2} \approx 0.3935$

$e^{-n/2} \le 0.01 \Rightarrow n \ge 10$

15. $f_X(x) = \dfrac{2}{\pi}\sqrt{1 - x^2}$, $-1 \le x \le 1$

$f_Y(y) = \dfrac{2}{\pi}\sqrt{1 - y^2}$, $-1 \le y \le 1$.

(a) Yes, $\mathbb{E}XY = \mathbb{E}X = \mathbb{E}Y = 0$.
(b) No, $f_{XY} \neq f_X f_Y$.

16. They both have the common ch.f. $\left(1 - \frac{iu}{\lambda}\right)^{-rn}$.

17. (a) (i) $\Phi(\frac{1}{2}) - \Phi(-\frac{1}{4}) \approx 0.2902$.
(ii) $\frac{1}{4} - \Phi^2(-\frac{3}{2}) \approx 0.2455$.
(b) (i) $e^{-4}(4^2/2! + 4^3/3! + 4^4/4!) \approx 0.537$.
(ii) $(1 - 3e^{-2})^2 - (1 - 7e^{-2})^2 \approx 0.350$.
(c) (i) $2.256(.8)^6 \approx 0.5914$.
(ii) $[1 - 1.8(.8)^4]^2 - (.2)^{10} \approx 0.0690$.
(d) (i) 11/16.
(ii) 63/256.
(e) (i) 13/18.
(ii) 4/9.

18. $$\int_0^\infty \int_{r_2}^\infty \frac{r_1 r_2}{a_1^2 a_2^2} e^{-(1/2)r_1^2/a_1^2} e^{-(1/2)r_2^2/a_2^2}\, dr_1\, dr_2 = \frac{a_1^2}{a_1^2 + a_2^2}$$

Section III

1. (a) The ch.f. of $\dfrac{X - \lambda}{\sqrt{\lambda}}$ is

$$\varphi_\lambda(u) = \exp[\lambda(e^{i(u/\sqrt{\lambda})} - 1) - iu\sqrt{\lambda}]$$

and $\lim_{\lambda \to \infty} \varphi_\lambda(u) = e^{-(1/2)u^2}$.

(b) The ch.f. of $\dfrac{\beta X - \alpha}{\sqrt{\alpha}}$ is

$$\varphi_\alpha(u) = \left(1 - \frac{iu}{\sqrt{\alpha}}\right)^{-\alpha} e^{-iu\sqrt{\alpha}}$$

and $\lim_{\alpha \to \infty} \varphi_\alpha(u) = e^{-(1/2)u^2}$.

2. Let $S_n = X_1 + \cdots + X_n$ where the X_is are i.i.d. Poisson(1). Then by the CLT

$$e^{-n} \sum_{k=0}^{n} \frac{n^k}{k!} = \mathbb{P}(S_n \le n) \to \frac{1}{2}.$$

3. $2\Phi(-3.8) \approx 0.00014$.

4. (a) $\Phi(-7.834) \approx 0$.
(b) $\Phi(2.655) \approx 0.9960$.
$\Phi(0.065) \approx 0.5259$.
$\Phi(-2.525) \approx 0.0058$.

Section IV

1. (a) $L_R = \{0, 4\}, L_T = \{1, 2, 3\}$

$$\rho = \begin{bmatrix} 1 & 0 & 0 & 0 & 0 \\ 7/15 & 2/7 & 2/3 & 4/7 & 8/15 \\ 3/15 & 3/7 & 4/9 & 6/7 & 12/15 \\ 1/15 & 1/7 & 1/3 & 2/7 & 14/15 \\ 0 & 0 & 0 & 0 & 1 \end{bmatrix}$$

$$m = \begin{bmatrix} 1 & \infty & \infty & \infty & \infty \\ \infty & \infty & \infty & \infty & \infty \\ \infty & \infty & \infty & \infty & \infty \\ \infty & \infty & \infty & \infty & \infty \\ \infty & \infty & \infty & \infty & 1 \end{bmatrix}$$

$$\mathbb{E}_x T_{L_R} = (17/5\ 18/5\ 11/5) \qquad x = 1, 2, 3.$$

(b) $L_R = \{0, 4\}, L_T = \{1, 2, 3\}$

$$\rho = \begin{bmatrix} 1 & 0 & 0 & 0 & 0 \\ 7/15 & 9/14 & 2/3 & 4/7 & 8/15 \\ 3/15 & 3/7 & 13/8 & 6/7 & 12/15 \\ 1/15 & 1/7 & 1/3 & 9/14 & 14/15 \\ 0 & 0 & 0 & 0 & 1 \end{bmatrix}$$

$$m = \begin{bmatrix} 1 & \infty & \infty & \infty & \infty \\ \infty & \infty & \infty & \infty & \infty \\ \infty & \infty & \infty & \infty & \infty \\ \infty & \infty & \infty & \infty & \infty \\ \infty & \infty & \infty & \infty & 1 \end{bmatrix}$$

$$\mathbb{E}_x T_{L_R} = (34/5\ 36/5\ 22/5) \qquad x = 1, 2, 3.$$

(c) $L_R = \{0\}, L_T = \{1, 2, 3, 4\}$

$$\rho = \begin{bmatrix} 1 & 0 & 0 & 0 & 0 \\ 1 & 2/3 & 2/3 & 4/7 & 8/15 \\ 1 & 1 & 8/9 & 6/7 & 12/15 \\ 1 & 1 & 1 & 20/21 & 14/15 \\ 1 & 1 & 1 & 1 & 44/45 \end{bmatrix}$$

$$m = \begin{bmatrix} 1 & \infty & \infty & \infty & \infty \\ 45 & \infty & \infty & \infty & \infty \\ 66 & 21 & \infty & \infty & \infty \\ 75 & 30 & 9 & \infty & \infty \\ 78 & 33 & 12 & 3 & \infty \end{bmatrix}$$

$$\mathbb{E}_x T_{L_R} = (45\ 66\ 75\ 78) \qquad x = 1, 2, 3, 4.$$

(d) $L_R = \{4\}, L_T = \{0, 1, 2, 3\}$

$$\rho = \begin{bmatrix} 29/45 & 1 & 1 & 1 & 1 \\ 7/15 & 13/21 & 1 & 1 & 1 \\ 3/15 & 3/7 & 5/9 & 1 & 1 \\ 1/15 & 1/7 & 1/3 & 1/3 & 1 \\ 0 & 0 & 0 & 0 & 1 \end{bmatrix}$$

$$m = \begin{bmatrix} \infty & 3/2 & 15/4 & 51/8 & 147/16 \\ \infty & \infty & 9/4 & 39/8 & 123/16 \\ \infty & \infty & \infty & 21/8 & 87/16 \\ \infty & \infty & \infty & \infty & 45/16 \\ \infty & \infty & \infty & \infty & 1 \end{bmatrix}$$

$$\mathbb{E}_x T_{L_R} = (147/16\ 123/16\ 87/16\ 45/16) \qquad x = 0, 1, 2, 3.$$

(e) $L_R = \{0, 4\}, L_T = \{1, 2, 3\}$

$$\rho = \begin{bmatrix} 1 & 0 & 0 & 0 & 0 \\ 3/4 & 1/3 & 1/2 & 1/3 & 1/4 \\ 1/2 & 2/3 & 3/4 & 2/3 & 1/2 \\ 1/4 & 1/3 & 1/2 & 5/6 & 3/4 \\ 0 & 0 & 0 & 0 & 1 \end{bmatrix}$$

$$m = \begin{bmatrix} 1 & \infty & \infty & \infty & \infty \\ \infty & \infty & \infty & \infty & \infty \\ \infty & \infty & \infty & \infty & \infty \\ \infty & \infty & \infty & \infty & \infty \\ \infty & \infty & \infty & \infty & 1 \end{bmatrix}$$

$$\mathbb{E} T_x L_R = (11/2\ 9\ 17/2) \qquad x = 1, 2, 3.$$

(f) $L_R = \{4\}, L_T = \{0, 1, 2, 3\}$

$$\rho = \begin{bmatrix} 7/8 & 1 & 1 & 1 & 1 \\ 3/4 & 5/6 & 1 & 1 & 1 \\ 1/2 & 2/3 & 7/8 & 1 & 1 \\ 1/4 & 1/3 & 1/2 & 7/8 & 1 \\ 0 & 0 & 0 & 0 & 1 \end{bmatrix}$$

$$m = \begin{bmatrix} \infty & 2 & 6 & 14 & 30 \\ \infty & \infty & 4 & 12 & 28 \\ \infty & \infty & \infty & 8 & 24 \\ \infty & \infty & \infty & \infty & 16 \\ \infty & \infty & \infty & \infty & 1 \end{bmatrix}$$

$$\mathbb{E} T_x L_R = (30\ 28\ 24\ 16) \qquad x = 0, 1, 2, 3.$$

(g) $L_R = \{0, 1, 2, 3, 4\}$

$$\rho = \begin{bmatrix} 1 & 1 & 1 & 1 & 1 \\ 1 & 1 & 1 & 1 & 1 \\ 1 & 1 & 1 & 1 & 1 \\ 1 & 1 & 1 & 1 & 1 \\ 1 & 1 & 1 & 1 & 1 \end{bmatrix}$$

$$m = \begin{bmatrix} 31 & 3/2 & 15/4 & 51/8 & 147/16 \\ 45 & 31/2 & 9/4 & 39/8 & 123/16 \\ 66 & 21 & 31/4 & 21/8 & 87/16 \\ 75 & 30 & 9 & 31/8 & 45/16 \\ 78 & 33 & 12 & 3 & 31/16 \end{bmatrix}.$$

(h) $L_R = \{0, 1, 2, 3, 4\}$

$$\rho = \begin{bmatrix} 1 & 1 & 1 & 1 & 1 \\ 1 & 1 & 1 & 1 & 1 \\ 1 & 1 & 1 & 1 & 1 \\ 1 & 1 & 1 & 1 & 1 \\ 1 & 1 & 1 & 1 & 1 \end{bmatrix}$$

$$m = \begin{bmatrix} 5 & 78/31 & 141/31 & 174/31 & 147/31 \\ 147/31 & 5 & 78/31 & 141/31 & 174/31 \\ 174/31 & 147/31 & 5 & 78/31 & 141/31 \\ 141/31 & 174/31 & 147/31 & 5 & 78/31 \\ 78/31 & 141/31 & 174/31 & 147/31 & 5 \end{bmatrix}.$$

(i) $L_R = \{0, 1\}, L_T = \{2, 3, 4, 5\}$

$$\rho = \begin{bmatrix} 1 & 0 & 0 & 0 & 0 & 0 \\ 0 & 1 & 0 & 0 & 0 & 0 \\ 3/4 & 1/4 & 5/8 & 1/4 & 1/8 & 1/2 \\ 1/4 & 3/4 & 1/4 & 5/8 & 1/8 & 1/2 \\ 1/2 & 1/2 & 1/2 & 1/2 & 1/4 & 1 \\ 1/2 & 1/2 & 1/2 & 1/2 & 1/4 & 5/8 \end{bmatrix}$$

$$m = \begin{bmatrix} 1 & \infty & \infty & \infty & \infty & \infty \\ \infty & 1 & \infty & \infty & \infty & \infty \\ \infty & \infty & \infty & \infty & \infty & \infty \\ \infty & \infty & \infty & \infty & \infty & \infty \\ \infty & \infty & \infty & \infty & \infty & 1 \\ \infty & \infty & \infty & \infty & \infty & \infty \end{bmatrix}$$

$$\mathbb{E}_x T_{L_R} = (29/6\ 29/6\ 20/3\ 17/3) \qquad x = 2, 3, 4, 5.$$

(j) $L_R = \{0, 1, 2, 4\}$, $L_T = \{3, 5\}$

$$\rho = \begin{bmatrix} 1 & 1 & 0 & 0 & 0 & 0 \\ 1 & 1 & 0 & 0 & 0 & 0 \\ 0 & 0 & 1 & 0 & 1 & 0 \\ 7/11 & 7/11 & 4/11 & 1/12 & 4/11 & 1/4 \\ 0 & 0 & 1 & 0 & 1 & 0 \\ 6/11 & 6/11 & 5/11 & 1/3 & 5/11 & 1/4 \end{bmatrix}$$

$$m = \begin{bmatrix} 5/2 & 2 & \infty & \infty & \infty & \infty \\ 3 & 5/3 & \infty & \infty & \infty & \infty \\ \infty & \infty & 13/6 & \infty & 8/7 & \infty \\ \infty & \infty & \infty & \infty & \infty & \infty \\ \infty & \infty & 4/3 & \infty & 13/7 & \infty \\ \infty & \infty & \infty & \infty & \infty & \infty \end{bmatrix}$$

$$\mathbb{E}_x T_{L_R} = (17/11\ 24/11) \qquad x = 3, 5.$$

(k) $L_R = \{0, 1, 2\}$, $L_T = \{3, 4, 5, 6\}$

$$\rho = \begin{bmatrix} 1 & 1 & 1 & 0 & 0 & 0 & 0 \\ 1 & 1 & 1 & 0 & 0 & 0 & 0 \\ 1 & 1 & 1 & 0 & 0 & 0 & 0 \\ 1 & 1 & 1 & 5/8 & 1/4 & 1/8 & 1/2 \\ 1 & 1 & 1 & 1/4 & 5/8 & 1/8 & 1/2 \\ 1 & 1 & 1 & 1/2 & 1/2 & 1/4 & 1 \\ 1 & 1 & 1 & 1/2 & 1/2 & 1/4 & 5/8 \end{bmatrix}$$

$$m = \begin{bmatrix} 3 & 3 & 3 & \infty & \infty & \infty & \infty \\ 3 & 3 & 3 & \infty & \infty & \infty & \infty \\ 3 & 3 & 3 & \infty & \infty & \infty & \infty \\ 13/2 & 67/12 & 85/12 & \infty & \infty & \infty & \infty \\ 13/2 & 85/12 & 67/12 & \infty & \infty & \infty & \infty \\ 7 & 49/6 & 49/6 & \infty & \infty & \infty & 1 \\ 6 & 43/6 & 43/6 & \infty & \infty & \infty & \infty \end{bmatrix}$$

$$\mathbb{E}_x T_{L_R} = (29/6\ 29/6\ 20/3\ 17/3) \qquad x = 3, 4, 5, 6.$$

(l) $L_R = \{1, 2, 3, 4, 5, 6\}$, $L_T = \{0\}$

$$\rho = \begin{bmatrix} \frac{1}{2} & \frac{3}{4} & \frac{3}{4} & \frac{3}{4} & \frac{1}{4} & \frac{1}{4} & \frac{1}{4} \\ 0 & 1 & 1 & 1 & 0 & 0 & 0 \\ 0 & 1 & 1 & 1 & 0 & 0 & 0 \\ 0 & 1 & 1 & 1 & 0 & 0 & 0 \\ 0 & 0 & 0 & 0 & 1 & 1 & 1 \\ 0 & 0 & 0 & 0 & 1 & 1 & 1 \\ 0 & 0 & 0 & 0 & 1 & 1 & 1 \end{bmatrix}$$

$$m = \begin{bmatrix} \infty & \infty & \infty & \infty & \infty & \infty & \infty \\ \infty & 3 & 1 & 2 & \infty & \infty & \infty \\ \infty & 2 & 3 & 1 & \infty & \infty & \infty \\ \infty & 1 & 2 & 3 & \infty & \infty & \infty \\ \infty & \infty & \infty & \infty & 3 & 4 & 2 \\ \infty & \infty & \infty & \infty & 2 & 3 & 4 \\ \infty & \infty & \infty & \infty & 4 & 2 & 3 \end{bmatrix}$$

$$\mathbb{E}_0 T_{L_R} = 2.$$

2. $$P(x, y) = \frac{\binom{2x}{y}\binom{2(d-x)}{d-y}}{\binom{2d}{d}},$$

$$\max(0, 2x - d) \le y \le \min(d, 2x)$$

This is hypergeometric with parameters $r = 2d$, $r_1 = 2x$, $r_2 = 2(d - x)$. Thus

$$\mathbb{E}(X_{n+1} | X_0, \ldots, X_n) = X_n.$$

(It's a martingale.) For $d = 5$

$$P = \begin{bmatrix} 1 & 0 & 0 & 0 & 0 & 0 \\ 2/9 & 5/9 & 2/9 & 0 & 0 & 0 \\ 1/42 & 10/42 & 20/42 & 10/42 & 1/42 & 0 \\ 0 & 1/42 & 10/42 & 20/42 & 10/42 & 1/42 \\ 0 & 0 & 0 & 2/9 & 5/9 & 2/9 \\ 0 & 0 & 0 & 0 & 0 & 1 \end{bmatrix}.$$

3. $$P(x, y) = \begin{cases} \dfrac{x^2}{d^2}, & y = x - 1 \\ 2\dfrac{x(d-x)}{d^2}, & y = x \\ \dfrac{(d-x)^2}{d^2}, & y = x + 1 \end{cases}$$

$$P = \begin{bmatrix} 0 & 1 & 0 & 0 & 0 & 0 \\ 1/25 & 8/25 & 16/25 & 0 & 0 & 0 \\ 0 & 4/25 & 12/25 & 9/25 & 0 & 0 \\ 0 & 0 & 9/25 & 12/25 & 4/25 & 0 \\ 0 & 0 & 0 & 16/25 & 8/25 & 1/25 \\ 0 & 0 & 0 & 0 & 1 & 0 \end{bmatrix}.$$

4. $L_R = \{0, 1, 2, \ldots\}$, $L_T = \phi$

$$\mathbb{P}_0(T_0 = n) = p^{n-1}q$$

5. $$\frac{\sum_{n=0}^{\infty} P^n(x, y)}{\sum_{n=0}^{\infty} P^n(y, y)} = \frac{\mathbb{E}_x N(y)}{1 + \mathbb{E}_y N(y)} = \rho_{xy}, \qquad \text{for } x \neq y.$$

6. For fixed x, $P(x, y)$ is binomial with parameters d and $\frac{x}{d}$. Thus $\mathbb{E}(X_{n+1} | X_0, \ldots, X_n) = X_n$. In particular $\mathbb{E}_x X_n = x$, $\forall n$. Thus for $T = T_{L_R}$

$$x = \mathbb{E}_x X_T = (0)\rho_{\{0\}}(x) + d\rho_{\{d\}}(x).$$

Therefore $\rho_{\{0\}}(x) = \dfrac{d - x}{d}$.

7. $\mathbb{E}_1 T_1 = 6$, $\mathbb{E}_1 T_1^2 = 76$

$\mathbb{E}_1 T_7 = \mathbb{E}_1 T_7^2 = \infty$.

8. $$P = \begin{pmatrix} 1 & 0 & 0 & 0 & 0 & 0 \\ 1/4 & 1/2 & 1/4 & 0 & 0 & 0 \\ 1/16 & 1/4 & 1/4 & 1/4 & 1/16 & 1/8 \\ 0 & 0 & 1/4 & 1/2 & 1/4 & 0 \\ 0 & 0 & 0 & 0 & 1 & 0 \\ 0 & 0 & 1 & 0 & 0 & 0 \end{pmatrix}.$$

9. $$P^m = 6^{-m} \begin{pmatrix} 1 & 2^m - 1 & 3^m - 2^m & 4^m - 3^m & 5^m - 4^m & 6^m - 5^m \\ 0 & 2^m & 3^m - 2^m & 4^m - 3^m & 5^m - 4^m & 6^m - 5^m \\ 0 & 0 & 3^m & 4^m - 3^m & 5^m - 4^m & 6^m - 5^m \\ 0 & 0 & 0 & 4^m & 5^m - 4^m & 6^m - 5^m \\ 0 & 0 & 0 & 0 & 5^m & 6^m - 5^m \\ 0 & 0 & 0 & 0 & 0 & 6^m \end{pmatrix}.$$

10. $$P = \begin{pmatrix} p & q & 0 & 0 \\ 0 & 0 & p & q \\ p & q & 0 & 0 \\ 0 & 0 & p & q \end{pmatrix}.$$

11. (a) 0.04796.
(b) \$8.32.

12. $\gamma_x = \dfrac{2}{(x + 1)(x + 2)}$. Thus the chain is transient.

(a) $\mathbb{P}_x(T_a < T_b) = \dfrac{a+1}{x+1}\dfrac{b-x}{b-a}.$

(b) $\rho_{x0} = \dfrac{1}{x+1},\ x \geq 1;\ \rho_{00} = \dfrac{1}{2}.$

13. $\gamma_x = \dfrac{1}{(x+1)^2}$. Thus the chain is transient.

$$\rho_{x0} = 1 - \frac{6}{\pi^2}\sum_{y=1}^{x}\frac{1}{y^2}, \quad x \geq 1; \quad \rho_{00} = \rho_{10}.$$

14. (a) 0.1.
(b) \$99.10.

15. $\mathbb{E}_x T_x = \dfrac{2^d}{\binom{d}{x}}$ (Check that $\pi(x) = \dfrac{\binom{d}{x}}{2^d}$.)

16. Same as above.

17. $\mathbb{E}_x T_x = \dfrac{\binom{2d}{d}}{\binom{d}{x}^2}.$

18. $\rho_{x0} = 1, \forall x.$

19. (a) $P^n(0,0) = \begin{cases} \binom{2m}{m} p^m q^m, & n = 2m, \\ 0, & n \text{ odd}. \end{cases}$

(b) $\displaystyle\sum_{n=0}^{\infty} P^n(0,0)x^n = (1 - 4pqx^2)^{-1/2}.$

(Use the fact that $\binom{2m}{m} = \binom{-\frac{1}{2}}{m}(-4)^m$.)

(c) $\displaystyle\mathbb{E}_0 x^{T_0} = 1 - \left[\sum_{n=0}^{\infty} P^n(0,0)x^n\right]^{-1}$

$= 1 - \sqrt{1 - 4pqx^2}$ (see page 103).

(d) $\rho_{00} = \lim_{x \uparrow 1} \mathbb{E}_0 x^{T_0} = 1 - \sqrt{1 - 4pq}.$

20. Set

$$\gamma(y) = \exp\left[-\int_a^y \frac{2v(z)}{\sigma^2(z)}\,dz\right].$$

Then

$$u(x) = \frac{\int_x^b \gamma(y)\,dy}{\int_a^b \gamma(y)\,dy},$$

$$m(x) = \frac{\int_a^x \gamma(y)\,dy}{\int_a^b \gamma(y)\,dy} \int_a^b \int_a^y \frac{\gamma(y)}{\gamma(z)}\,dz\,dy - \int_a^x \int_a^y \frac{\gamma(y)}{\gamma(z)}\,dz\,dy$$

(cf. with the formulas for u and v on pages 84–86).

21. $\rho = \sqrt{5} - 2.$

$\rho_{20} = (\sqrt{5} - 2)^2 = 9 - 4\sqrt{5}.$

22. $\rho = \min\left(\frac{p}{q}, 1\right) \quad \left(\text{solve } \frac{p}{1 - q\rho} = \rho\right)$

$\rho = \min\left(\frac{1 + p - \sqrt{q(1 + 3p)}}{2q}, 1\right) \quad \left(\text{solve } \left(\frac{p}{1 - q\rho}\right)^2 = \rho\right).$

23. The chain is irreducible if and only if $p_Z(0) > 0$, $p_Z(0) + p_Z(1) < 1$.

(a) $p_Z(0) = 0$, $p_Z(1) < 1$: $L_T = \{0, 1, 2, \ldots\}$
$p_Z(0) > 0$, $p_Z(0) + p_Z(1) = 1$: $L_R = \{0\}$, $L_T = \{1, 2, \ldots\}$.
(b) $p_Z(1) = 1$: $L_R = \{0, 1, 2, \ldots\}$.

24. The chain is recurrent if and only if

$$\sum_{n=1}^{\infty} q_n = \infty.$$

25. (a) $\mathbb{P}(T = k) = \binom{N-1}{k-2} \frac{(k-1)!}{N^{k-1}}, 2 \le k \le N + 1.$

(b) We could estimate N from

$$\mathbb{E}T = 2 + \sum_{k=1}^{N-1} \prod_{i=1}^{k} \left(1 - \frac{i}{N}\right).$$

26. Refer to condition (i) on page 116.

Consider the system $\sum_{y=1}^{\infty} P(x, y)u(y) = u(x)$, $x \ge 1$. Letting $U(t) = \sum_{x=1}^{\infty} u(x)t^x$ we find that

$$U(t) = p_Z(0)u(1)\frac{t}{\Phi_Z(t) - t}.$$

Now if $\mu = \mathbb{E}Z > 1$ then (see page 93) the generating function Φ_Z has a fixed point $t \in [0, 1)$, and thus U cannot have bounded coefficients (since it blows up at this fixed point). If $\mu \le 1$ then write

$$\Phi_Z(t) - t = (1 - t)[1 - \psi(t)],$$

where $\psi(t) = \sum_{x=0}^{\infty} \mathbb{P}(Z > x)t^x$. Thus if we write

$$\sum_{k=0}^{\infty} \psi^k(t) = \sum_{k<0}^{\infty} C_k t^k$$

then $C_k \ge 0$ and

$$U(t) = p_Z(0)u(1)\frac{t}{1-t}\sum_{k=0}^{\infty} \psi^k(t) = p_Z(0)u(1)t\sum_{n=0}^{\infty}\left(\sum_{k=0}^{n} C_k\right)t^n.$$

Thus the coefficients of U are bounded if and only if $\sum_k C_k < \infty$, which is the case if and only if $\mu = \psi(1) < 1$. Thus the chain is

$$\text{transient if } \mu < 1 \qquad \text{and} \qquad \text{recurrent if } \mu \ge 1.$$

27. Observe that

$$\mathbb{E}_x \varepsilon^{dX_1} = (\alpha_x \varepsilon^d + 1 - \alpha_x)^d = \varepsilon^{dx},$$

and so $\mathbb{E}_x \varepsilon^{dX_n} = \varepsilon^{dx}$, $\forall n$. For $T = T_{L_R}$

$$\varepsilon^{dx} = \mathbb{E}_x \varepsilon^{dX_T} = \rho_{\{0\}}(x) + \rho_{\{d\}}(x)\varepsilon^{d^2},$$

and so

$$\rho_{\{d\}}(x) = \frac{1 - \varepsilon^{dx}}{1 - \varepsilon^{d^2}}.$$

28. $m_{10} = m_{21} = m_{32} = m_{03} = \dfrac{1 + 2p^2}{p^2 + q^2}.$

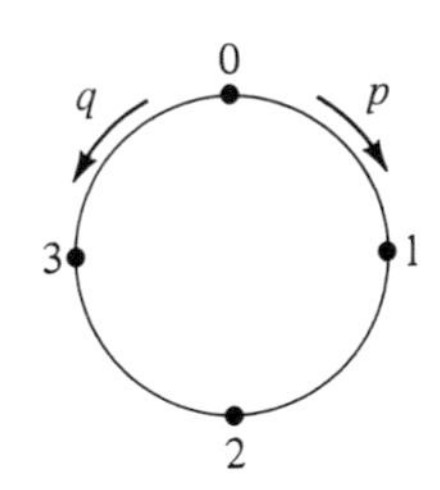

$m_{20} = m_{31} = m_{02} = m_{13} = \dfrac{2}{p^2 + q^2}.$

$m_{30} = m_{01} = m_{12} = m_{23} = \dfrac{1 + 2q^2}{p^2 + q^2}.$

$m_{00} = m_{11} = m_{22} = m_{33} = 4.$

29. By the formula for the linear chain on page 88,

$$m_{x0} = m_{x+1\,1} = \cdots = m_{x-1\,d}$$

$$= \frac{x}{q-p} - \frac{d+1}{q-p}\frac{(q/p)^x - 1}{(q/p)^{d+1} - 1}, \qquad x \ne 0,$$

and $m_{xx} = d + 1$.

30. $$P^n(\mathbf{0},\mathbf{0}) = \begin{cases} \binom{2m}{m}^2 \Big/ 4^{2m}, & n = 2m, \\ 0, & n \text{ odd}. \end{cases}$$

By Stirling's formula, $P^{2m}(\mathbf{0},\mathbf{0}) \sim \dfrac{1}{\pi m}$, so this random walk is recurrent. For a 3-D random walk

$$P^n(\mathbf{0},\mathbf{0}) = \begin{cases} \dfrac{\binom{2m}{m}}{6^{2m}} \sum_{i+j \le m} \binom{m}{i,\, j,\, m-i-j}^2, & n = 2m, \\ 0, & n \text{ odd}. \end{cases}$$

It can be shown as in Karlin and Taylor [31] that $\sum P^n(\mathbf{0},\mathbf{0}) < \infty$, so that this random walk is transient.

Section V

1. (a) $$\lim_n \frac{1}{n} \sum_{m=1}^{n} P^m = \begin{bmatrix} 1 & 0 & 0 & 0 & 0 \\ 7/15 & 0 & 0 & 0 & 8/15 \\ 3/15 & 0 & 0 & 0 & 12/15 \\ 1/15 & 0 & 0 & 0 & 14/15 \\ 0 & 0 & 0 & 0 & 1 \end{bmatrix}$$

$\lim_n P^n =$ same.

(b) Same as (a).

(c) $$\lim_n \frac{1}{n} \sum_{m=1}^{n} P^m = \begin{bmatrix} 1 & 0 & 0 & 0 & 0 \\ 1 & 0 & 0 & 0 & 0 \\ 1 & 0 & 0 & 0 & 0 \\ 1 & 0 & 0 & 0 & 0 \\ 1 & 0 & 0 & 0 & 0 \end{bmatrix}$$

$\lim_n P^n =$ same.

(d) $$\lim_n \frac{1}{n} \sum_{m=1}^{n} P^m = \begin{bmatrix} 0 & 0 & 0 & 0 & 1 \\ 0 & 0 & 0 & 0 & 1 \\ 0 & 0 & 0 & 0 & 1 \\ 0 & 0 & 0 & 0 & 1 \\ 0 & 0 & 0 & 0 & 1 \end{bmatrix}$$

$\lim_n P^n =$ same.

(e) $$\lim_n \frac{1}{n}\sum_{m=1}^{n} P^m = \begin{pmatrix} 1 & 0 & 0 & 0 & 0 \\ 3/4 & 0 & 0 & 0 & 1/4 \\ 1/2 & 0 & 0 & 0 & 1/2 \\ 1/4 & 0 & 0 & 0 & 3/4 \\ 0 & 0 & 0 & 0 & 1 \end{pmatrix}$$

$\lim_n P^n =$ same.

(f) Same as (d).

(g) $$\lim_n \frac{1}{n}\sum_{m=1}^{n} P^m = \begin{pmatrix} \frac{1}{31} & \frac{2}{31} & \frac{4}{31} & \frac{8}{31} & \frac{16}{31} \\ \frac{1}{31} & \frac{2}{31} & \frac{4}{31} & \frac{8}{31} & \frac{16}{31} \\ \frac{1}{31} & \frac{2}{31} & \frac{4}{31} & \frac{8}{31} & \frac{16}{31} \\ \frac{1}{31} & \frac{2}{31} & \frac{4}{31} & \frac{8}{31} & \frac{16}{31} \\ \frac{1}{31} & \frac{2}{31} & \frac{4}{31} & \frac{8}{31} & \frac{16}{31} \end{pmatrix}$$

$\lim_n P^n =$ same.

(h) $$\lim_n \frac{1}{n}\sum_{m=1}^{n} P^m = \begin{pmatrix} \frac{1}{5} & \frac{1}{5} & \frac{1}{5} & \frac{1}{5} & \frac{1}{5} \\ \frac{1}{5} & \frac{1}{5} & \frac{1}{5} & \frac{1}{5} & \frac{1}{5} \\ \frac{1}{5} & \frac{1}{5} & \frac{1}{5} & \frac{1}{5} & \frac{1}{5} \\ \frac{1}{5} & \frac{1}{5} & \frac{1}{5} & \frac{1}{5} & \frac{1}{5} \\ \frac{1}{5} & \frac{1}{5} & \frac{1}{5} & \frac{1}{5} & \frac{1}{5} \end{pmatrix}$$

$\lim_n P^n =$ same.

(i) $$\lim_n \frac{1}{n}\sum_{m=1}^{n} P^m = \begin{pmatrix} 1 & 0 & 0 & 0 & 0 & 0 \\ 0 & 1 & 0 & 0 & 0 & 0 \\ 3/4 & 1/4 & 0 & 0 & 0 & 0 \\ 1/4 & 3/4 & 0 & 0 & 0 & 0 \\ 1/2 & 1/2 & 0 & 0 & 0 & 0 \\ 1/2 & 1/2 & 0 & 0 & 0 & 0 \end{pmatrix}$$

$\lim_n P^n =$ same.

(j) $$\lim_n \frac{1}{n}\sum_{m=1}^{n} P^m = \begin{pmatrix} 2/5 & 3/5 & 0 & 0 & 0 & 0 \\ 2/5 & 3/5 & 0 & 0 & 0 & 0 \\ 0 & 0 & 6/13 & 0 & 7/13 & 0 \\ 14/55 & 21/55 & 24/143 & 0 & 28/143 & 0 \\ 0 & 0 & 6/13 & 0 & 7/13 & 0 \\ 12/55 & 18/55 & 30/143 & 0 & 35/143 & 0 \end{pmatrix}$$

$\lim_n P^n =$ same.

(k) $$\lim_n \frac{1}{n}\sum_{m=1}^{n} P^m = \begin{bmatrix} \frac{1}{3} & \frac{1}{3} & \frac{1}{3} & 0 & 0 & 0 & 0 \\ \frac{1}{3} & \frac{1}{3} & \frac{1}{3} & 0 & 0 & 0 & 0 \\ \frac{1}{3} & \frac{1}{3} & \frac{1}{3} & 0 & 0 & 0 & 0 \\ \frac{1}{3} & \frac{1}{3} & \frac{1}{3} & 0 & 0 & 0 & 0 \\ \frac{1}{3} & \frac{1}{3} & \frac{1}{3} & 0 & 0 & 0 & 0 \\ \frac{1}{3} & \frac{1}{3} & \frac{1}{3} & 0 & 0 & 0 & 0 \\ \frac{1}{3} & \frac{1}{3} & \frac{1}{3} & 0 & 0 & 0 & 0 \end{bmatrix}$$

$\lim_n P^n =$ same.

(l) $$\lim_n \frac{1}{n}\sum_{m=1}^{n} P^m = \begin{bmatrix} 0 & \frac{1}{4} & \frac{1}{4} & \frac{1}{4} & \frac{1}{12} & \frac{1}{12} & \frac{1}{12} \\ 0 & \frac{1}{3} & \frac{1}{3} & \frac{1}{3} & 0 & 0 & 0 \\ 0 & \frac{1}{3} & \frac{1}{3} & \frac{1}{3} & 0 & 0 & 0 \\ 0 & \frac{1}{3} & \frac{1}{3} & \frac{1}{3} & 0 & 0 & 0 \\ 0 & 0 & 0 & 0 & \frac{1}{3} & \frac{1}{3} & \frac{1}{3} \\ 0 & 0 & 0 & 0 & \frac{1}{3} & \frac{1}{3} & \frac{1}{3} \\ 0 & 0 & 0 & 0 & \frac{1}{3} & \frac{1}{3} & \frac{1}{3} \end{bmatrix}$$

$$\lim_n P^{3n} = \begin{bmatrix} 0 & \frac{8}{28} & \frac{9}{28} & \frac{4}{28} & \frac{1}{12} & \frac{1}{12} & \frac{1}{12} \\ 0 & 1 & 0 & 0 & 0 & 0 & 0 \\ 0 & 0 & 1 & 0 & 0 & 0 & 0 \\ 0 & 0 & 0 & 1 & 0 & 0 & 0 \\ 0 & 0 & 0 & 0 & \frac{1}{3} & \frac{1}{3} & \frac{1}{3} \\ 0 & 0 & 0 & 0 & \frac{1}{3} & \frac{1}{3} & \frac{1}{3} \\ 0 & 0 & 0 & 0 & \frac{1}{3} & \frac{1}{3} & \frac{1}{3} \end{bmatrix}$$

$$\lim_n P^{3n+1} = \begin{bmatrix} 0 & \frac{4}{28} & \frac{8}{28} & \frac{9}{28} & \frac{1}{12} & \frac{1}{12} & \frac{1}{12} \\ 0 & 0 & 1 & 0 & 0 & 0 & 0 \\ 0 & 0 & 0 & 1 & 0 & 0 & 0 \\ 0 & 1 & 0 & 0 & 0 & 0 & 0 \\ 0 & 0 & 0 & 0 & \frac{1}{3} & \frac{1}{3} & \frac{1}{3} \\ 0 & 0 & 0 & 0 & \frac{1}{3} & \frac{1}{3} & \frac{1}{3} \\ 0 & 0 & 0 & 0 & \frac{1}{3} & \frac{1}{3} & \frac{1}{3} \end{bmatrix}$$

$$\lim_n P^{3n+2} = \begin{bmatrix} 0 & \frac{9}{28} & \frac{4}{28} & \frac{8}{28} & \frac{1}{12} & \frac{1}{12} & \frac{1}{12} \\ 0 & 0 & 0 & 1 & 0 & 0 & 0 \\ 0 & 1 & 0 & 0 & 0 & 0 & 0 \\ 0 & 0 & 1 & 0 & 0 & 0 & 0 \\ 0 & 0 & 0 & 0 & \frac{1}{3} & \frac{1}{3} & \frac{1}{3} \\ 0 & 0 & 0 & 0 & \frac{1}{3} & \frac{1}{3} & \frac{1}{3} \\ 0 & 0 & 0 & 0 & \frac{1}{3} & \frac{1}{3} & \frac{1}{3} \end{bmatrix}.$$

2. $\pi(x) = \begin{cases} \dfrac{1}{2c}, & 1 \le x \le c, \\ \dfrac{1}{2d}, & c + 1 \le x \le c + d. \end{cases}$

3. Suppose the chain is in fact positive recurrent. Then

$$\frac{1}{n} \sum_{m=1}^{n} P^m(x, y) \to \pi(y), \qquad \forall x, y.$$

Let $L_1 \subseteq L$ be a finite subset of the state space. Summing both sides over $x \in L_1$ and using the fact that P^m is doubly stochastic,

$$1 \ge \frac{1}{n} \sum_{m=1}^{n} \sum_{x \in L_1} P^m(x, y) \to \pi(y)|L_1|.$$

Thus $\pi(y) \le \dfrac{1}{|L_1|}$ and since $|L_1|$ is unbounded we arrive at the contradiction $\pi(y) = 0$.

4. $\pi(x) = p^x q,\ x \ge 0.$

5. (a) Choose n so that $P^n(x, y) > 0$. Then

$$\pi(y) = \sum_z \pi(z) P^n(z, y) \ge \pi(x) P^n(x, y) > 0.$$

(b) $\pi(y) = \sum_x \pi(x) P(x, y) = c \sum_x \pi(x) P(x, z) = c\pi(z).$

6. $\pi(x) = \dfrac{1}{(x + 1)!(e - 1)}$. Yes, it is positive recurrent. No, the second case is not positive recurrent, but it is null recurrent.

7. Since $y \to x$ we can choose k so that $P^k(y, x) > 0$. Then $P^{k+m}(y, y)$ and $P^{k+n}(y, y)$ are both positive. Thus $d_y | k + m$ and $d_y | k + n$, and so $d_y | n - m$.

8. Define a (lower triangular) transition probability matrix

$$Q(x, y) = \beta_y \prod_{k=y+1}^{x} (1 - \beta_k), \qquad 0 \le y \le x.$$

Then in terms of Q we have

$$P(x, y) = \begin{cases} \left(1 - \dfrac{\beta_{x+1}}{\beta_x}\right) Q(x, y), & 0 \le y \le x, \\ \dfrac{\beta_{x+1}}{\beta_x}, & y = x + 1. \end{cases}$$

We claim that the nth row of Q is the first row of P^n. Clearly this holds if $n = 0$. For $n > 0$ use induction:

$$\sum_y Q(n, y)P(y, z) = \sum_{y=z}^{n} \left(1 - \frac{\beta_{y+1}}{\beta_y}\right) Q(n, y)Q(y, z) + \frac{\beta_z}{\beta_{z-1}} Q(n, z-1)$$

$$= \sum_{y=z}^{n} (\beta_y - \beta_{y+1}) Q(n, z) + \frac{\beta_z}{\beta_{z-1}} Q(n, z-1)$$

$$= (\beta_z - \beta_{n+1}) Q(n, z) + \frac{\beta_z}{\beta_{z-1}} Q(n, z-1)$$

$$= Q(n+1, z).$$

Thus in particular $P^n(0, 0) = Q(n, 0) = \left(\sum_{k=0}^{n} b_k\right)^{-1}$. Since $\sum_{k=0}^{n} b_k \leq n + 1$ we have $\sum_n P^n(0, 0) = \infty$, so our chain is recurrent. Since the chain is aperiodic, we conclude that it is

$$\text{positive recurrent if } \sum_k b_k < \infty,$$

$$\text{null recurrent if } \sum_k b_k = \infty.$$

In the former case, the stationary distribution is given by

$$\pi(y) = \lim_x Q(x, y) = \frac{\sum_{k=0}^{y} b_k}{\sum_{k=0}^{\infty} b_k} \beta_y.$$

9. $F(x) = Px$ where P is doubly stochastic. Thus, since the chain is aperiodic,

$$\lim_n F^n(w) = (\overline{w}, \ldots, \overline{w}),$$

where $\overline{w} = \frac{1}{2k+1} \sum_i w_i$.

10. By Jensen's inequality

$$\varphi(P^{k+1}(x, y)) = \varphi\left(\sum_z P(x, z)P^k(z, y)\right) \geq \sum_z P(x, z)\varphi(P^k(z, y)).$$

Thus

$$\sum_x \pi(x)\varphi(P^{k+1}(x, y)) \geq \sum_x \sum_z \pi(x)P(x, z)\varphi(P^k(z, y))$$

$$= \sum_z \pi(z)\varphi(P^k(z, y)).$$

11. In general $\mathbb{P}(X_0 = y | X_1 = x) = \frac{\pi(y)}{\pi(x)} P(y, x)$. Thus for the birth and death chain

$$\mathbb{P}(X_0 = y | X_1 = x) = \begin{cases} q_x, & y = x - 1 \\ r_x, & y = x \\ p_x, & y = x + 1 \end{cases}$$

which turns out to be simply $P(x, y)$.

12. $$\pi(x) = \begin{cases} \frac{1}{2}\left(1 - \frac{p}{q}\right), & x = 0 \\ \frac{1}{2}\left(1 - \frac{p}{q}\right)\frac{p^{x-1}}{q^x}, & x \geq 1. \end{cases}$$

13. Here $p_x = 1 - \frac{x}{d}$ and $q_x = \frac{x}{d}$. The stationary distribution is binomial with parameters d and $1/2$. Thus its mean and variance are $\frac{1}{2}d$ and $\frac{1}{4}d$, respectively. (See also Problem 15 in Chapter IV.)

14. For $n > m$

$$\mathbb{E}(X_n | X_m = x) = \frac{\lambda}{q}(1 - p^{n-m}) + p^{n-m}x.$$

Thus

$$\begin{aligned} \mathbb{E}X_m X_n &= \mathbb{E}\mathbb{E}(X_m X_n | X_m) \\ &= \mathbb{E}\left[\frac{\lambda}{q}(1 - p^{n-m})X_m + p^{n-m}X_m^2\right] = \frac{\lambda}{q}\left(\frac{\lambda}{q} + p^{n-m}\right). \end{aligned}$$

In particular $\operatorname{Cov}(X_m, X_n) = \frac{\lambda}{q} p^{n-m}$.

15. $\pi(x) = p^x q$, $x \geq 0$. (cf. Problem 4).

16. $\mathbb{P}_0(T_0 > n) \leq \sum_{x=n}^{\infty} p_x$ and this tends to zero as $n \to \infty$. Thus the chain is always recurrent. It follows inductively that

$$m_{x0} = \sum_{k=1}^{x} \frac{1}{k}, \qquad x \geq 1.$$

Thus $m_{00} = 1 + \sum_{x=1}^{\infty} p_x m_{x0}$, and so the chain is positive recurrent if and only if $\sum_x p_x \log x < \infty$. In this case

$$\pi(x) = \frac{1}{K}\left[p_x + \frac{1}{x+1}\left(1 - \sum_{y=0}^{x} p_y\right)\right],$$

where $K = 1 + \sum_{x=1}^{\infty} p_x\left(\sum_{y=1}^{x} \frac{1}{y}\right)$.

17. $\pi = (\frac{1}{3}\ \frac{1}{9}\ \frac{2}{9}\ \frac{1}{12}\ \frac{1}{4})$ and the period is 3.

18.
$$\lim_n P^{2n} = \begin{pmatrix} \frac{1}{8} & 0 & \frac{6}{8} & 0 & \frac{1}{8} \\ 0 & \frac{4}{8} & 0 & \frac{4}{8} & 0 \\ \frac{1}{8} & 0 & \frac{6}{8} & 0 & \frac{1}{8} \\ 0 & \frac{4}{8} & 0 & \frac{4}{8} & 0 \\ \frac{1}{8} & 0 & \frac{6}{8} & 0 & \frac{1}{8} \end{pmatrix}$$

$$\lim_n P^{2n+1} = \begin{pmatrix} 0 & \frac{4}{8} & 0 & \frac{4}{8} & 0 \\ \frac{1}{8} & 0 & \frac{6}{8} & 0 & \frac{1}{8} \\ 0 & \frac{4}{8} & 0 & \frac{4}{8} & 0 \\ \frac{1}{8} & 0 & \frac{6}{8} & 0 & \frac{1}{8} \\ 0 & \frac{4}{8} & 0 & \frac{4}{8} & 0 \end{pmatrix}.$$

19. $\pi = (\frac{3}{11}\ \frac{1}{11}\ \frac{1}{11}\ \frac{6}{11})$. Thus a typical sequence is CCCSSSSCCCCC. Thus $\frac{4}{11}$ of the days are sunny in the long run.

20. (a)

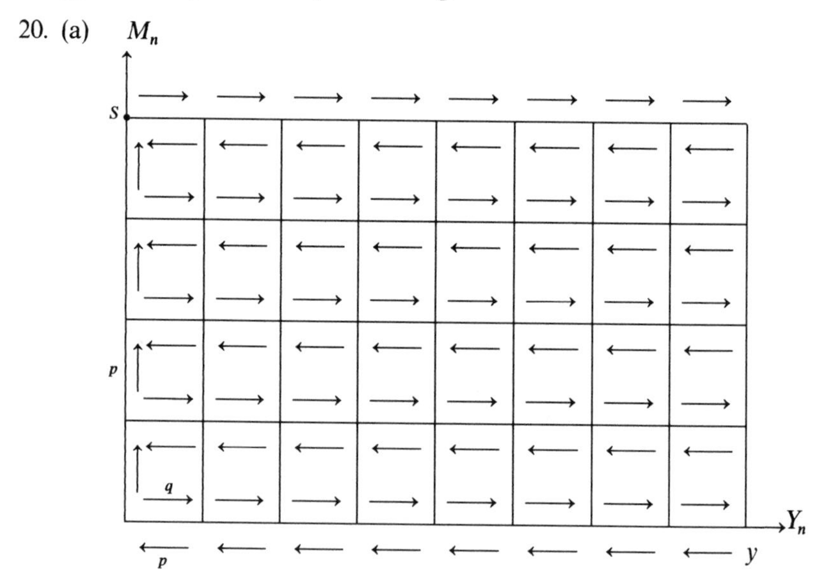

In order for the random walk (M_n, Y_n) to exit at the point $(s, 0)$ before reaching the right wall $Y_n = y$, we must have $\mathbb{P}_0(T_{-1} < T_y)$ for the gambler's ruin with probabilities p (right) and q (left), at each level $M_n = m$. Thus (see page 88)

$$\mathbb{P}\left(\max_{0 \le k \le T(s)} Y_k < y\right) = \begin{cases} \left[\dfrac{\rho(p - (q/p)^{y+1}}{1 - (q/p)^{y+1}}\right]^s, & p \neq q, \\ \left(\dfrac{y}{y+1}\right)^s, & p = q. \end{cases}$$

(b) $\{M_{\tau(y)} \geq a\} = \left\{\max_{0 \le k \le T(0)} Y_k < y\right\}$. Thus θ is given by the preceding expression.

Index

A
Abel's Lemma, 102
Absorbing state, 83, 84, 88
Absorbtion probability, 82
Almost sure convergence, 45
Attractor, 165

B
Backward equation, 124
Barnsley, M. F., 167
Bellman, R., 166
Bernoulli distribution, 8
Berry–Esseen Theorem, 53, 56
Beta distribution, 19
Billingsley, P., 165
Binomial distribution, 9, 26
Birkhoff, G., 141
Birth and death chain, 84ff, 98, 118, 119
  aperiodic, 113
  mean first passage times, 85, 86
  null recurrent, 113
  periodic, 113
  positive recurrent, 113
  transient, 85, 113
Birth and death process, 125, 129ff
  examples, 130–135
  linear, 134, 135
  Poisson, 127ff
  pure birth, 126
  pure death, 126
  rates, 125
Bochner's Theorem, 57ff
Borel–Cantelli Lemma, 46–48
Branching chain, 91–93, 99
  extinction probability, 91
Branching process, 133, 134
  with immigration, 134
Breiman, L., 166
Brownian motion, 90

C
Cantor function, 21, 22
Cauchy distribution, 19, 25, 65
Cauchy–Schwarz Inequality, 29
Central Limit Theorem, 52, 77, 115, 116, 149
  multi-variate, 57, 58
Chapman–Kolmogorov equation, 123
Characteristic function, 4, 14, 23, 26, 29, 41, 169
  conditional, 30
  inversion formula, 5, 15, 23, 42
  joint, 29, 31, 41
  properties of, 6, 14
Chebyshev's Inequality, 8
Chi-square distribution, 18

Closed set of states (for a Markov chain), 80
Computer generation (of random variables), 23
  multiplicative congruential method, 24
Conditional
  characteristic function, 30
  conditional probability density, 29
  expectation, 30, 33ff
  variance, 30
Cohen, J. E., 167
Collage property, 159, 167, 169
Continuity Theorem, 51, 57
Contractivity
  average, 159
  for operator, 142
Convergence
  almost sure, 45
  density function, 11, 27, 30, 40
  in distribution, 48
  in probability, 45, 52, 61, 63
  weak, 48ff
Convex
  function, 3, 42, 69
  hull, 169, 171
  set, 36
Convolution formula, 28, 31
Correlation coefficient, 29
Covariance, 28, 41
  matrix, 38, 41
Cramer's Theorem, 70ff
  examples, 75
Cumulant, 7

D

Density function, 11, 27, 30, 40
  conditional, 29
  discrete, 1
  joint continuous, 30
  joint discrete, 27
  marginal, 27
Diffusion coefficient (of a diffusion process), 89
Diffusion process (as limit of Markov chain), 89–91
Distributions
  Bernoulli, 8
  beta, 18
  binomial, 9
  Cauchy, 18, 65, 149
  chi-square, 18
  exponential, 17
  F-, 18
  gamma, 17
  geometric, 10
  hypergeometric, 9, 10
  Laplace, 25
  lognormal, 25
  Maxwell, 25
  negative binomial, 10, 11
  normal, 16, 17, 25, 26, 38, 39, 42, 52, 53, 57, 65
  Pareto, 64, 65
  Pascal, 25
  Poisson, 11
  Polya, 25
  Rayleigh, 18
  t-, 18
  uniform, 16, 23–25, 65
  z-, 18
Distribution function (cumulative), 11, 20ff
  absolutely continuous, 20
  absolutely continuous part of, 21
  continuous part of, 20
  degenerate, 65
  discrete, 20
  discrete part of, 20
  extremal, 67
  joint, 30
  multivariate, 39ff
  nondegenerate, 65
  singular continuous, 21
  singular part of, 21
  sub-, 48
Dominated Convergence Theorem, 5
Drift coefficient (of a diffusion process), 89

E

Ehrenfest chain, 99, 112, 119
  diffusion limit of, 90
  modified, 99, 112, 113
Ellis, R. S., 70
Ergodic
  geometric ergodicity (for a Markov chain), 110

Ergodic (*cont.*)
process, 140
theorems, 140ff
Erlang's Loss Formula, 132
Estimation
linear, 37
nonlinear, 37
Expectation, 1, 12, 23, 28
properties of, 2, 3, 13
Exponential distribution, 17, 25
Exterior product, 150
Extinction probability (for a branching chain), 91
Extremal distribution function, 67
Extreme point, 169
Extremes, 60ff, 63

F
F-distribution, 18
First passage, 80
mean times, 83
probabilities, 80, 84, 89
Fisher, R., 18
Forward equation, 124
Fourier series, 4, 5
Fractal, 156ff
random algorithm for generating, 156, 166–169
Fubini's Theorem, 5
Furstenberg, H., 167

G
Gambler's ruin, 86, 88, 98
Gamma distribution, 17
Generating function
for a branching chain, 91–93
for a Markov chain, 102–103
for a queueing chain, 94, 95
moment, 7, 69
probability, 7, 35
Geometric distribution, 10
Gosset, W. S. ("Student"), 19

H
Hammersley, J. M., 166
Helly–Bray Theorem, 49, 57
Herglotz Lemma, 58
Hilbert space, 36, 37
Hitting time, 80, 125
Holder Inequality, 32, 69
Hypergeometric distribution, 9, 10

I
Independent increments, 127
Independent random variables, 27, 30, 40
Interarrival time (for a queue), 93
Invariant
event, 140, 146
minimal, 158
set, 158
Inversion formula, 5, 15, 23
Irreducible (set of states), 80
Iterated function system, 167

J
Jensen's Inequality, 3, 42, 69

K
Kendall, M., 93
Kesten, H., 166
Khinchine's Convergence of Types Theorem, 65, 66
Kingman, J. F. C., 143, 167
Koch snowflake curve, 171
Kolmogorov
Inequality, 46
Zero-One Law, 48, 140
Krengel, U., 166–167
Kullback–Liebler information, 75

L
Laplace distribution, 25
Law of Large Numbers, 45ff
Strong, 101, 115, 149, 162
strong form, 45, 47, 48, 101, 115, 149, 162
weak form, 45, 51, 52, 69, 71
Law of Types Theorem, 67–69
Leads to (for states of a Markov chain), 80

Legendre–Fenchel transform, 69
Lognormal distribution, 25

M

Marginal density, 27
Markov chain, 78ff
  aperiodic, 107
  birth and death, 84ff
  branching chain, 91–93
  examples, 111ff
  generator of, 84–86, 89, 90, 116
  hitting time for, 80
  initial distribution of, 79
  irreducible, 80
  null recurrent, 103
  passage probabilities, 80
  period of, 107
  periodic, 107
  positive recurrent, 103
  queueing chain, 93–96
  random walk, 86
  recurrent, 80
  stationary, 78
  stationary distribution of, 101ff, 104
  transient, 80
  two-state, 111
Markov process, 121ff
  birth and death, 125ff
  embedded chain of, 95, 124
  explosive, 121, 137
  generator of, 123
  hitting time for, 125
  infinitesimal parameters of, 124
  irreducible, 125
  jump times, 121
  null recurrent, 125
  passage probabilities, 125
  Poisson, 127ff
  positive recurrent, 125
  pure, 121
  rates of, 121
  recurrent, 125
  stationary distribution of, 125
Markov property, 78, 122
Maximal Ergodic Theorem, 140
Maxwell distribution, 25
Mean Ergodic Theorem, 142
Minimax Theorem, 71, 73
Moment, 4, 41
  central, 4
Moment generating function, 7

N

Negative binomial distribution, 10, 11
Newman, C. M., 167
Normal distribution, 16, 17, 25, 26, 52, 53, 65
  joint, 38, 39, 42, 57
Null recurrent state (of a Markov chain), 103

O

Ornstein-Uhlenbeck process, 90
Orthogonal projection, 36
Oseledec's Theorem, 149ff
  example, 155

P

Pareto distribution, 64, 65
Pascal distribution, 25
Point mass, 20
Pointwise Ergodic Theorem, 140–141
Poisson distribution, 11, 25
  approximation to binomial, 52
Poisson process, 125ff
  compound, 128
  filtered, 128
  point process, 129, 137, 138
Polya distribution, 25
Positive recurrent state (of a Markov chain), 103
Prohorov's Theorem, 49, 57

Q

Queueing chain, 93–96, 99, 100
  embedded chain of queueing process, 95, 96
  GI/D/1, 94, 95, 113, 114
  GI/M/1 (embedded chain of), 95, 96
  M/G/1 (embedded chain of), 95

Queueing process, 130–133
customer loss ratio, 132
infinite server, 133
M/M/N, 130
utilization factor, 132
waiting time distribution, 131

R
Raghunathan, M. S., 167
Random variable
absolutely continuous, 11
compound, 35
discrete, 1ff
joint absolutely continuous, 30–32
joint discrete, 27–30
Random walk, 86, 99, 100
gambler's ruin, 86, 88
simple, 86–89
Rate function, 70
Rayleigh distribution, 19
Recurrent state (of a Markov chain), 80
Riemann–Stieltjes integral, 22, 23
Rockafellar, R. T., 69

S
Self-covering set, 158
Semi-group, 123
generator of, 123
Markov, 124
Service time (for a queue), 93
Simulation, 23, 24
Smoothing Inequality, 53
Spectral radius (for matrix distribution), 148
Standard deviation, 4
Stationary
asymptotic, 162
distribution (for Markov chain), 101ff, 104
distribution (for Markov process), 125
distribution (with respect to random matrix), 157
ergodic, 140
increments, 127
process, 140
Stirling's approximation, 56, 75
Subadditive Ergodic Theorem, 143ff
Subadditive process, 143
Decomposition Theorem for, 144
ergodic, 146
time constant of, 144
Subdistribution function, 48
Support, 157

T
t-distribution, 19
Tail event, 48
Taylor's Theorem, 6, 62
Tight sequence (of distribution functions), 48
Transformation of random variables, 12, 31
Transient state (of a Markov chain), 80
Transition probability, 78, 121

U
Uniform distribution, 16, 23, 25, 65
simulation of, 24

V
Variance, 4

W
Waiting time, 101, 121
Weak convergence, 48ff
Welsh, J. A. D., 166

Y
Yule process, 134

Z
z-distribution, 18

Springer Texts in Statistics *(continued from p. ii)*

Keyfitz	Applied Mathematical Demography Second Edition
Kiefer	Introduction to Statistical Inference
Kokoska and Nevison	Statistical Tables and Formulae
Lindman	Analysis of Variance in Experimental Design
Madansky	Prescriptions for Working Statisticians
McPherson	Statistics in Scientific Investigation: Its Basis, Application, and Interpretation
Nguyen and Rogers	Fundamentals of Mathematical Statistics: Volume I: Probability for Statistics
Nguyen and Rogers	Fundamentals of Mathematical Statistics: Volume II: Statistical Inference
Noether	Introduction to Statistics: The Nonparametric Way
Peters	Counting for Something: Statistical Principles and Personalities
Pfeiffer	Probability for Applications
Santner and Duffy	The Statistical Analysis of Discrete Data
Saville and Wood	Statistical Methods: The Geometric Approach
Sen and Srivastava	Regression Analysis: Theory, Methods, and Applications
Whittle	Probability via Expectation, Third Edition
Zacks	Introduction to Reliability Analysis: Probability Models and Statistical Methods